RAND NATIONAL DEFENSE RESEARCH INSTITUTE

Counterinsurgency Scorecard Update

Afghanistan in Early 2015 Relative to Insurgencies Since World War II

Christopher Paul, Colin P. Clarke

Prepared for the Office of the Secretary of Defense

For more information on this publication, visit www.rand.org/t/RR1273

Library of Congress Cataloging-in-Publication Data is available for this publication.
ISBN: 978-0-8330-9262-5

Published by the RAND Corporation, Santa Monica, Calif.

Cover: U.S. Army photo by Pfc. David Devich.

Preface

This short report seeks to replicate and update similar RAND efforts published in 2011 and 2013, *Counterinsurgency Scorecard: Afghanistan in Early 2011 Relative to the Insurgencies of the Past 30 Years* and *Counterinsurgency Scorecard: Afghanistan in Early 2013 Relative to Insurgencies Since World War II*, respectively. The core of the research involved conducting an expert elicitation exercise (using classic Delphi methods) to complete the counterinsurgency (COIN) scorecard developed as part of other earlier studies. The COIN scorecard was first published in 2010 as part of *Victory Has a Thousand Fathers: Sources of Success in Counterinsurgency* and then updated in the 2013 report *Paths to Victory: Lessons from Modern Insurgencies*. The expert elicitation for this update was completed in March and April 2015.

The goal of this effort was to once again apply the methods developed through this earlier research to examine historical insurgencies and assess the progress of ongoing COIN efforts in Afghanistan as of early 2015.

This work will be of interest to defense analysts and military planners who are responsible for evaluating current U.S. operations and COIN approaches, particularly in Afghanistan; to military and civilian decisionmakers with responsibility for Afghanistan; to academics and scholars who engage in historical research on COIN, insurgency, and irregular warfare; and to students of contemporary and historic international conflicts.

This research was conducted within the International Defense and Security Policy Center of the RAND National Defense Research Institute, a federally funded research and development center sponsored by the Office of the Secretary of Defense, the Joint Staff, the Unified Combatant Commands, the Navy, the Marine Corps, the defense agencies, and the defense Intelligence Community.

For more information on the International Security and Defense Policy Center, see www.rand.org/nsrd/ndri/centers/isdp or contact the director (contact information is provided on the web page).

Contents

Figure and Tables

Figure

Tables

Summary

The RAND report *Paths to Victory: Lessons from Modern Insurgencies* used detailed case studies of 71 insurgencies worldwide since World War II to identify correlates of success in counterinsurgency (COIN). One of the core findings of that effort was that historical governments' scores on a scorecard of 15 good factors or practices minus 11 bad factors or practices corresponded with the outcomes of all of the 59 core cases considered.[1] Those outcomes were either "COIN win," which included cases in which the government unambiguously prevailed and cases in which the government had the better of a mixed outcome, or "COIN loss," cases in which the insurgents clearly prevailed or had the better of a mixed outcome. Cases in which COIN forces were able to maximize the presence of good factors and minimize the presence of bad factors resulted in COIN force success. Specifically, cases with a good-minus-bad score of +2 or greater were always won by the government, and cases with a good-minus-bad score of –1 or lower were always won by the insurgents.[2] The research found that all successful COIN forces have a scorecard score of at least +2 *and* that all successful COIN campaigns realized three specific factors: the disruption of tangible support to the insurgents, the demonstration of commitment and motivation on the part of both the government and COIN forces, and flexibility and adaptability on the part of COIN forces.

The current research effort involved conducting an expert elicitation exercise to complete the scorecard for operations in Afghanistan as of early 2015 (when Afghan security forces were unambiguously the primary counterinsurgent actor, with some residual support from international partners). It is the third in a series of similar exercises to the scorecard for Afghanistan, with the previous two exercises completed in 2011 and 2013.[3]

We asked a panel of 14 experts on Afghanistan to make "worst-case" assessments of the scorecard factors. According to the consensus results, 7.5 good factors and 5.5 bad factors were

[1] The study included a total of 71 cases, but that full case set included instances of COIN campaigns fought "against the tide of history"—for example, at the end of the colonial period or apartheid—and one case of indeterminate outcome. This left the 59 core cases, which were used for the analyses and became the foundation for the scorecard.

[2] Perfect discrimination of 59 historical cases is by no means a guarantee of perfect prediction across all possible cases. Full details of the historical research can be found in Christopher Paul, Colin P. Clarke, Beth Grill, and Molly Dunigan, *Paths to Victory: Lessons from Modern Insurgencies*, Santa Monica, Calif.: RAND Corporation, RR-291/1-OSD, 2013b.

[3] Christopher Paul, *Counterinsurgency Scorecard: Afghanistan in Early 2011 Relative to the Insurgencies of the Past 30 Years*, Santa Monica, Calif.: RAND Corporation, OP-337-OSD, 2011; Christopher Paul, Colin P. Clarke, Beth Grill, and Molly Dunigan, *Counterinsurgency Scorecard: Afghanistan in Early 2013 Relative to Insurgencies Since World War II*, Santa Monica, Calif.: RAND Corporation, RR-396-OSD, 2013a.

present in early 2015 Afghanistan, for an overall score of +2.[4] This represents no change in the overall score from the 2013 expert elicitation, though there were slight changes in the specific scorecard factors found to be present or absent. In short, progress was measured with two steps forward, two steps back. The experts identified improvements from previous scorecards in government legitimacy and good governance factors, most likely the result of renewed optimism surrounding the recently elected administration of Ashraf Ghani and the promise many felt in the departure of Ghani's predecessor, the mercurial Hamid Karzai. But now that the honeymoon with Ghani is ending, it appears that Afghanistan is left with many of the same unresolved issues as when the conflict began, including corruption, poor governance, and weak rule of law. If the scorecard exercise were repeated at the time of this report's publication, the expert panel might not be so optimistic about the extent of improvement under the Ghani administration.

The overall score of +2 for both 2013 and 2015 represents an improvement over the score generated through a similar process in early 2011. Afghanistan's total score of +2 is tied with the lowest-scoring winning historical COIN forces, and, with 7.5 good factors, is consonant with historical COIN success. However, only historical cases in which the insurgents prevailed had more than five bad factors—and early 2015 Afghanistan had 5.5. Also concerning is the repeated observation by participants that many of the positive factors identified in early 2015 Afghanistan would be difficult for the Afghans to maintain after most international forces withdraw.

Despite an overall score tied with historical winners, the analysis does not suggest that a strictly military victory by the government of Afghanistan is very likely, due to the absence of (and slim prospects for achieving) several critical factors. The best prospects for the conflict's resolution are through some kind of negotiated settlement, and future U.S. efforts should focus on reaching such a settlement and on favorable terms.

This report highlights two critical shortcomings, one of which—the failure to disrupt tangible support to the insurgents—has plagued efforts by COIN forces since the very beginning of the campaign to subdue the insurgency. Cross-border sanctuary in Pakistan remains available, allowing insurgent groups like the Haqqani Network to continue to receive safe haven as they plan and organize attacks on Afghan security forces and coalition troops. Also identified in both earlier reports and still significant matters of concern are issues like corruption, arbitrary personalistic rule (persistent subnational rule by strong power brokers and patronage networks, with capricious decisions by those with authority), and the perverse incentives of elites to perpetuate the conflict, as well as Afghanistan's dependence on external supporters.

The following factors were assessed as absent in contemporary Afghanistan but found to be essential to success in historical COIN campaigns: (1) disrupting tangible support to the insurgents and (2) a demonstration (and improvement) of commitment and motivation on the part of the Afghan government and Afghan security forces. The absence of this latter factor—the commitment and motivation of the Afghan government and security forces—has become even more worrisome in light of the partial collapse of Iraqi security forces in the face of the Islamic State threat. Finally, these same two factors were also absent in 2013 and 2011, and both were identified as critical to the COIN campaign then. Their absence remained a major

[4] "Half" factors indicate that the expert panel could not reach consensus. Any factor present was scored "1," any factor absent was scored "0"; factors on which the panel could not reach consensus were scored as neither present nor absent, 0.5.

concern in 2015 and even more so following the continued gradual drawdown of international support personnel.

COIN forces in Afghanistan are still unable to disrupt the tangible support of the insurgents, and the experts discerned that there had been only slight improvement in the commitment and motivation of the COIN force. Still, improvement in other areas, including consistency with message, increased legitimacy, and improved unity of effort could very well have spillover benefits for the commitment and motivation of the force if these gains are real and can be properly channeled. None of these changes (positive or negative) are inevitable, though, and as the mission in Afghanistan has demonstrated repeatedly over the past 14 years of conflict, gains are hard-won and easily reversed.

Based on these results and the results of other research (our 2014 study *From Stalemate to Settlement: Lessons for Afghanistan from Historical Insurgencies That Have Been Resolved Through Negotiations*), we recommend that, going forward, U.S. support for Afghanistan emphasize efforts to increase the prospects for a successful negotiated settlement. This will be especially effective where such efforts are "dual use" in that they contribute both to a path to settlement and to improving the Afghan government's negotiating position by improving its military position. Since the first requirement of negotiated settlement is military stalemate, efforts to support Afghan government forces and diminish the support available to insurgents serve both ends. Also, U.S. efforts to pressure external actors should not just emphasize reduced military support for the insurgents, but also increased support for negotiations and settlement. To avoid a repeat of similar missteps in Iraq, the United States should push for reconciliation to ensure an inclusive negotiated settlement (likely to include constitutional reform) that gives the Taliban a legitimate voice in the political process. By achieving this objective, U.S. policy would work to drive a wedge between the internally directed insurgency of the Taliban and the transnational threat posed by groups like al Qaeda and the Islamic State.[5]

---

[5] Jeff Eggers, "Afghanistan, Choose Your Enemies Wisely," *Foreign Policy*, August 24, 2015.

Counterinsurgency Scorecard Update: Afghanistan in Early 2015 Relative to Insurgencies Since World War II

Findings from Previous Research on Insurgency

The RAND report *Paths to Victory: Lessons from Modern Insurgencies* used detailed case studies of the 71 insurgencies begun and completed worldwide between World War II and 2010 to identify and analyze correlates of success in counterinsurgency (COIN).[1] "Success" in this research denotes unambiguous government victory or a mixed outcome that favored the government. Cases in which the government unambiguously prevailed or a mixed outcome favored the government were coded as a "win"; any other outcome was coded as a "loss." This is relevant to 2015 Afghanistan in that, while there appears to be broad consensus that an unambiguous military victory for Afghan security forces is likely out of reach, some kind of mixed result that is more or less favorable to the Afghan government may not be. *Paths to Victory* produced several key findings regarding success in COIN in modern history that might be relevant for Afghanistan:

- Effective COIN practices run in packs, and some practices are always in the pack: tangible support reduction (reducing flows of insurgent funding, manpower, weapons, and intelligence and the availability of safe havens), a clearly committed and motivated COIN force, and a flexible and adaptable COIN force.
- Every insurgency is unique, but not so much that it matters at this level of analysis; the COIN scorecard discriminates all cases into wins and losses.
- The quality of COIN forces is more important than the quantity, especially where paramilitaries and irregular forces are concerned.
- Once good COIN practices are in place, the average insurgency lasts roughly six more years.
- Poor beginnings do not necessarily lead to poor ends.

Of particular interest is the second part of the second key finding, "The COIN scorecard discriminates cases into wins and losses." The COIN scorecard was originally developed as part of an earlier study, findings from which were published in the 2010 RAND report *Victory Has*

[1] Note that while the study included 71 detailed case studies, only 59 of these cases are considered "core" cases and informed the quantitative and comparative analyses. The excluded cases were "fought against the tide of history"—that is, cases in which the outcome was all but predetermined by exogenous global trends, such as the end of colonialism or the end of apartheid. Many of those cases were individually interesting and informative but made poor comparative cases; regardless of how well-designed and -executed these COIN campaigns may have been, they ultimately failed because of inexorable changes in the context of the conflict. Full details can be found in Christopher Paul, Colin P. Clarke, Beth Grill, and Molly Dunigan, *Paths to Victory: Lessons from Modern Insurgencies*, Santa Monica, Calif.: RAND Corporation, RR-291/1-OSD, 2013b.

a Thousand Fathers: Sources of Success in Counterinsurgency.[2] We applied that original COIN scorecard to Afghanistan in early 2011.[3] In 2013, we updated the scorecard based on new data compiled in support of the *Paths to Victory* study. Extending that work, we focus here on the importance of maximizing good factors and minimizing bad factors in defeating insurgencies, as well as how COIN forces might move toward a favorable negotiated political settlement, assessing the balance of these factors in Afghanistan in early 2015.[4]

The Counterinsurgency Scorecard

As noted, one of the core findings of the previous efforts was that a historical COIN case's score on a scorecard of 15 equally weighted good factors or practices minus 11 equally weighted bad factors or practices completely discriminated case outcomes in the data. Table 1 lists these good and bad factors.

Across the 59 core cases in the *Paths to Victory* analyses, taking the sum of the good minus the bad revealed that cases with a good-minus-bad score of +2 or greater were always won by the government, and cases with a good-minus-bad score of –1 or lower were always won by the insurgents.[5] In other words, scores on the scorecard completely discriminated the historical cases into wins and losses.[6] The lesson for current or future counterinsurgents is clear: Strategies that maximize the presence of good factors and minimize the presence of bad factors are endorsed by history.

Table 2 lists the 59 core insurgencies from *Paths to Victory*, along with the dates of the conflict, the sum of good factors (of a possible 15), the sum of bad factors (of a possible 11), the net of good-minus-bad factors, and the outcome of the case (either a COIN loss or a COIN win).[7] Note that the scorecard score does indeed discriminate the wins from the losses, with all winning COIN forces scoring at least +2 and all losing COIN forces scoring –1 or lower.

Endgames for Insurgencies: Getting to Negotiated Settlements

In follow-on research to *Paths to Victory* we considered an additional question related to the conclusions of mixed-outcome insurgencies: How do internal wars get to conflict-terminating negotiated settlements? Results from those analyses are presented in *From Stalemate to*

---

[2] Christopher Paul, Colin P. Clarke, and Beth Grill, *Victory Has a Thousand Fathers: Sources of Success in Counterinsurgency*, Santa Monica, Calif.: RAND Corporation, MG-964-OSD, 2010.

[3] See Christopher Paul, *Counterinsurgency Scorecard: Afghanistan in Early 2011 Relative to the Insurgencies of the Past 30 Years*, Santa Monica, Calif.: RAND Corporation, OP-337-OSD, 2011. For background on the scorecard's development, see Paul, Clarke, and Grill, 2010.

[4] Our thanks to quality assurance reviewers Jeff Eggers and Daniel Egel for their helpful comments on an earlier draft of this report.

[5] Note, again, that *win* refers to either clear victory or getting the better of a mixed outcome, while *loss* refers to either insurgent success or a mixed outcomes favoring the insurgents. Full details can be found in Paul, Clarke, Grill, and Dunigan, 2013b.

[6] Note that perfect discrimination in these 59 cases does not guarantee perfect discrimination across all possible future cases. With a larger sample size (additional cases and data), it is possible—even likely—that the estimated model would change slightly. In fact, this occurred when the data set went from 30 cases (*Victory Has a Thousand Fathers*) to 59 cases (*Paths to Victory*).

[7] For full details on the case selection and data collection processes and the factor scoring for each case, see Paul, Clarke, Grill, and Dunigan, 2013b.

Table 1
Good and Bad COIN Factors and Practices

15 Good COIN Practices	11 Bad COIN Practices
The COIN force realized at least two strategic communication factors.	The COIN force used both collective punishment and escalating repression.
The COIN force reduced at least three tangible support factors.	There was corrupt and arbitrary personalistic government rule.
The government realized at least one government legitimacy factor.	Host-nation elites had perverse incentives to continue the conflict.
Government corruption was reduced/good governance increased since the onset of the conflict.	An external professional military engaged in fighting on behalf of the insurgents.
The COIN force realized at least one intelligence factor.	The host nation was economically dependent on external supporters.
The COIN force was of sufficient strength to force the insurgents to fight as guerrillas.	Fighting was initiated primarily by the insurgents.
Unity of effort/unity of command was maintained.	The COIN force failed to adapt to changes in adversary strategy, operations, or tactics.
The COIN force avoided excessive collateral damage, disproportionate use of force, or other illegitimate application of force.	The COIN force engaged in more coercion or intimidation than the insurgents.
The COIN force sought to engage and establish positive relations with the population in the area of conflict.	The insurgent force was individually superior to the COIN force by being either more professional or better motivated.
Short-term investments, improvements in infrastructure or development, or property reform occurred in the area of conflict controlled or claimed by the COIN force.	The COIN force or its allies relied on looting for sustainment.
The majority of the population in the area of conflict supported or favored the COIN force.	The COIN force and government had different goals or levels of commitment.
The COIN force established and then expanded secure areas.	
Government/COIN reconstruction/development sought/achieved improvements that were substantially above the historical baseline.	
The COIN force provided or ensured the provision of basic services in areas that it controlled or claimed to control.	
The perception of security was created or maintained among the population in areas that the COIN force claimed to control.	

Settlement: Lessons for Afghanistan from Historical Insurgencies That Have Been Resolved Through Negotiations.[8]

For that study, we began with the same 59 core cases listed in Table 2, but then considered only those that had a mixed outcome (each outcome favored one side or the other and so would still be considered a win or loss in Table 2). Each of those 13 conflicts involved a mixed outcome in which substantial concessions were made to the losing side, and each was resolved through some form of negotiated settlement. The 13 cases were Indonesia (Aceh), Northern Ireland, Bosnia, Chechnya, the Lebanese Civil War, Burundi, Kampuchea, Tajikistan, Western Sahara, Congo (anti-Kabila), Mozambique (RENAMO), Philippines (MNLF), and Yemen.

From these 13 cases, we developed a master narrative of the historical steps to resolving an insurgency through negotiated settlement. The master narrative for negotiated settlements unfolds following these seven steps generally, though not always in this exact order:

[8] Colin P. Clarke and Christopher Paul, *From Stalemate to Settlement: Lessons for Afghanistan from Historical Insurgencies That Have Been Resolved Through Negotiations*, Santa Monica, Calif.: RAND Corporation, RR-469-OSD, 2014.

Table 2
59 Core Cases and Scorecard Scores

Country (Insurgency)	Years	Good COIN Practices	Bad COIN Practices	Total Score	Outcome
South Vietnam	1960–1975	0	–11	–11	COIN loss
Somalia	1980–1991	0	–9	–9	COIN loss
Afghanistan (post-Soviet)	1992–1996	0	–9	–9	COIN loss
Kosovo	1996–1999	0	–8	–8	COIN loss
Liberia	1989–1997	1	–9	–8	COIN loss
Cambodia	1967–1975	0	–7	–7	COIN loss
Moldova	1990–1992	1	–8	–7	COIN loss
Georgia/Abkhazia	1992–1994	0	–7	–7	COIN loss
Zaire (anti-Mobutu)	1996–1997	0	–7	–7	COIN loss
Nicaragua (Somoza)	1978–1979	0	–6	–6	COIN loss
Chechnya I	1994–1996	1	–7	–6	COIN loss
Bosnia	1992–1995	0	–6	–6	COIN loss
Laos	1959–1975	2	–7	–5	COIN loss
Nagorno-Karabakh	1992–1994	0	–5	–5	COIN loss
Democratic Republic of the Congo (anti-Kabila)	1998–2003	1	–5	–4	COIN loss
Rwanda	1990–1994	2	–6	–4	COIN loss
Bangladesh	1971	2	–6	–4	COIN loss
Afghanistan (Taliban)	1996–2001	2	–6	–4	COIN loss
Kampuchea	1978–1992	0	–3	–3	COIN loss
Cuba	1956–1959	3	–6	–3	COIN loss
Eritrea	1961–1991	1	–4	–3	COIN loss
Sudan (Sudanese People's Liberation Army [SPLA])	1984–2004	1	–4	–3	COIN loss
Afghanistan (anti-Soviet)	1978–1992	2	–5	–3	COIN loss
Burundi	1993–2003	1	–3	–2	COIN loss
Yemen	1962–1970	1	–3	–2	COIN loss
Lebanese Civil War	1975–1990	5	–7	–2	COIN loss
Tajikistan	1992–1997	2	–4	–2	COIN loss
Nepal	1997–2006	3	–4	–1	COIN loss
Indonesia (East Timor)	1975–2000	3	–4	–1	COIN loss
Nicaragua (Contras)	1981–1990	3	–4	–1	COIN loss
Papua New Guinea	1988–1998	2	–3	–1	COIN loss
Iraqi Kurdistan	1961–1975	4	–2	2	**COIN win**
Western Sahara	1975–1991	4	–2	2	**COIN win**
Argentina	1969–1979	5	–2	3	**COIN win**
Oman (Imamate Uprising)	1957–1959	4	–1	3	**COIN win**
Croatia	1992–1995	5	–2	3	**COIN win**
Guatemala	1960–1996	8	–4	4	**COIN win**
Tibet	1956–1974	7	–3	4	**COIN win**
Sri Lanka	1976–2009	6	–1	5	**COIN win**
Mozambique (Mozambican National Resistance [RENAMO])	1976–1995	8	–3	5	**COIN win**

Table 2—Continued

Country (Insurgency)	Years	Good COIN Practices	Bad COIN Practices	Total Score	Outcome
Turkey (Kurdistan Workers' Party [PKK])	1984–1999	8	–2	6	**COIN win**
Indonesia (Aceh)	1976–2005	8	–2	6	**COIN win**
Algeria (Armed Islamic Group [GIA])	1992–2004	6	0	6	**COIN win**
Baluchistan	1973–1978	9	–2	7	**COIN win**
Uganda (Allied Democratic Forces [ADF])	1986–2000	7	0	7	**COIN win**
Northern Ireland	1969–1999	9	–1	8	**COIN win**
Jordan	1970–1971	9	0	9	**COIN win**
Indonesia (Darul Islam)	1958–1962	10	0	10	**COIN win**
Angola (National Union for the Total Independence of Angola [UNITA])	1975–2002	12	–2	10	**COIN win**
Greece	1945–1949	12	–2	10	**COIN win**
Uruguay	1963–1972	10	0	10	**COIN win**
Malaya	1948–1955	13	–2	11	**COIN win**
El Salvador	1979–1992	12	–1	11	**COIN win**
Oman (Dhofar Rebellion)	1965–1975	13	–1	12	**COIN win**
Peru	1980–1992	14	–2	12	**COIN win**
Sierra Leone	1991–2002	14	–1	13	**COIN win**
Senegal	1982–2002	13	0	13	**COIN win**
Philippines (Moro National Liberation Front [MNLF])	1971–1996	14	–1	13	**COIN win**
Philippines (Huk Rebellion)	1946–1956	15	0	15	**COIN win**

First, after years of fighting, both sides to the conflict reach a state of war weariness and settle into a mutually hurting military stalemate. Second, after a stalemate has been reached and the belligerents recognize the futility of continued escalation, the insurgents are accepted as a legitimate negotiating partner. Once the insurgents have been accepted by the government, the terms of a ceasefire can be discussed. This third step, ceasefire, like step 2 before it, is highly dependent on the acquiescence of external powers. For example, if an active external supporter is pushing for continued conflict, it is likely that the negotiation process will end here.

If external actors refrain from further meddling, official intermediate agreements can be reached, the fourth step in the narrative. Fifth, power-sharing offers (such as amnesty or elections) can further entice the insurgents to favor politics over armed struggle. Sixth, once offers of power-sharing have been accepted, a moderation of the insurgency's leadership can facilitate further progress by giving a voice to the politically minded cadre of the group. Seventh, and finally, third-party guarantors are required to help guide the process to a close, acting as impartial observers or providers of security, economic assistance, development aid, and other services.

Figure 1 shows the master narrative for insurgencies that progress from conflict to negotiated settlement. While not all of the cases considered unfolded exactly according to this sequence (in fact, only one did), each case unfolded in a manner close enough to this narrative that it is a useful comparative tool for understanding how to get to negotiated settlements, and it should prove useful in identifying a path to conflict resolution in Afghanistan.

Figure 1
Seven Steps to Negotiated Settlement

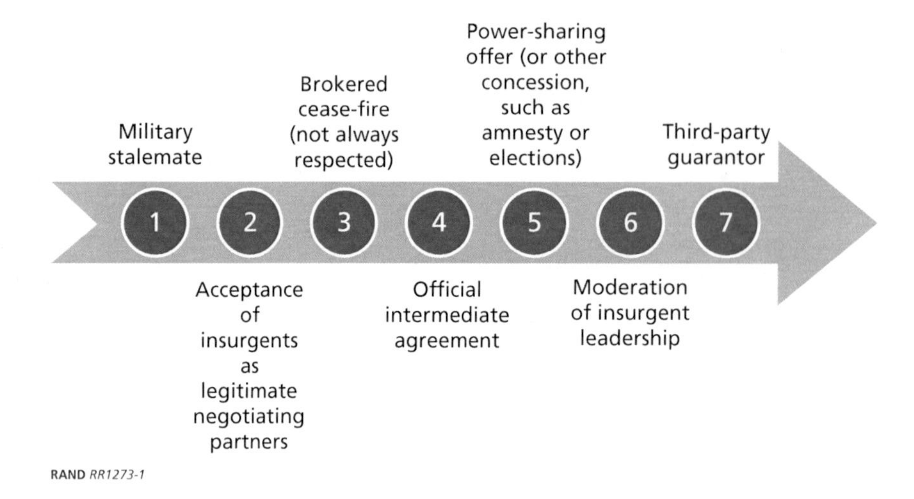

RAND *RR1273-1*

Scoring the Ongoing Conflict in Afghanistan

In addition to providing an interesting summary of historical results and reflecting historical correlates of success in COIN, the scorecard is meant to be a useful prognostication tool for current and future COIN efforts. Although we do not assert that the scorecard's ability to discriminate wins from losses since World War II guarantees its ability to predict the outcomes of ongoing or future conflicts, scores for these COIN efforts should offer a good indication of whether an effort is on a path likely to lead to a favorable outcome or on one that is less likely to do so. Specific factors scored as present or absent should also provide guidance to decision-makers, strategists, and planners about areas ripe for renewed emphasis.[9]

In the foundational historical research, we assessed the scorecard factors for each phase of each case based on detailed case studies compiled from numerous secondary sources. However, available data on Afghanistan often lack reliability and validity due to inconsistent collection practices, inconsistent reporting, and genuine ambiguity about what has happened in many parts of the country. Because the conflict is ongoing and current reporting often lacks the clarity of hindsight, we were unable to apply our historical methods with the same level of confidence. Instead, we employed a panel of experts to complete the scorecard for Afghanistan in early 2015 and elicited their input using Delphi methods. The next section describes the Delphi method; details specific to our implementation are provided in the appendix.

---

[9] Note that the scorecard is not meant to be a mindless checklist or substitute for a thoughtful and considered COIN strategy. Which factors can be achieved, how difficult they are to achieve, and the interrelationships between the factors all vary with the specific context, and every insurgency must be considered within its own context. The principal value of the scorecard is for diagnostic purposes. If a given COIN strategy is not producing a positive scorecard balance, careful scrutiny of the scorecard can suggest whether the strategy is on the right track (several good factors are likely under the strategy but have not yet been realized) or unlikely to lead to success (unlikely to lead to many good factors, especially those found to be essential to historical COIN wins).

Using the Delphi Method

The Delphi method was developed at RAND in the 1960s. While the technique has been refined over the years,[10] the fundamental premise remains the same. Experts individually make assessments or provide input and then offer written justification for those assessments. These experts are then given the opportunity to privately review the justifications offered by other participants and revise their assessments based on lines of reasoning that they had failed to include in their own initial calculations. The result is a consensual set of expert assessments based on more information than any single expert initially considered. Because participants work in private and remain anonymous to each other, final evaluations are reached without any of the psychological pitfalls of committee work, such as "specious persuasion, the unwillingness to abandon publicly expressed opinions, and the bandwagon effect of majority opinion."[11]

Still, it should be noted that the Delphi method is merely one of several methods that could be used to measure progress (or lack thereof) in Afghanistan or for populating the COIN scorecard. We might have populated the scorecard ourselves, relying on single-expert judgment and citing polls, interviews, and other reports or data sources to justify our scores. However, we felt that this would put undue weight on our personal opinions and views and potentially camouflage significant and interesting contentions and different points of view.

Because it can be difficult and contentious to identify specific scorecard factors as present or absent, we conducted an expert elicitation exercise with a panel of 14 subject-matter experts who were knowledgeable about ongoing operations in Afghanistan, keeping responses and participants anonymous. The iterative elicitation took place via email between March 11 and April 20, 2015. Panelists included RAND staff with expertise on Afghanistan or who had deployed to Afghanistan; serving field-grade U.S. military officers with multiple (and recent) deployments to Afghanistan; military veterans who were current on COIN research; defense civilian representatives from the Office of the Secretary of Defense, the Joint Staff, and U.S. Central Command; government civilians with experience in Afghanistan; staff from the U.S. Agency for International Development; university faculty members; journalists; and experts from other prominent think tanks. Full details of the expert elicitation process are provided in the appendix.

In keeping with the approach used in the foundational historical research used to generate the scorecard and the two previous Delphi-based scoring exercises for Afghanistan (held in 2011 and 2013), participants were asked to make *worst-case assessments*. That meant that to be considered present, positive factors were required to be present over the preponderance of the area of conflict, not just in isolated locations. Similarly, bad factors were counted as present if they occurred with any frequency greater than isolated incidents. So, good factors needed to be preponderantly present, but bad factors needed only to be routinely present in certain areas or among certain segments of the force. (Further details about the scoring instructions can be found in the appendix.)

---

[10] See, for example, Carolyn Wong, *How Will the e-Explosion Affect How We Do Research? Phase I: The E-DEL+I Proof-of-Concept Exercise*, Santa Monica, Calif.: RAND Corporation, DB-399-RC, 2003.

[11] Bernice B. Brown, *Delphi Process: A Methodology Used for the Elicitation of Opinions of Experts*, Santa Monica, Calif.: RAND Corporation, P-3925, 1968, p. 2.

Building on Previous Scorecards

Our first effort to employ the scorecard for the ongoing conflict in Afghanistan using Delphi methods came in 2011 with the initial report, *Counterinsurgency Scorecard: Afghanistan in Early 2011 Relative to the Insurgencies of the Past 30 Years.*[12] In 2011, experts scored several critical governance factors as absent, including "Government a functional democracy" and "Free and fair elections held." Two years after the 2011 scorecard, we repeated the expert elicitation process and produced a follow-up report, *Counterinsurgency Scorecard: Afghanistan in Early 2013 Relative to Insurgencies Since World War II.* The 2013 scorecard revealed that many of the same deficiencies in terms of governance were still present.[13] Governance factors scored as absent in 2013 included "Government leaders selected in a manner considered just and fair by majority of population in area of conflict" and "Majority of citizens in the area of conflict view government as legitimate." Without question, the absence of these critical governance factors was highly concerning to participating experts in both 2011 and 2013.

The inability of the COIN forces to disrupt the tangible support of the insurgents has been an enduring issue dating back to the very beginning of the current conflict in Afghanistan and was scored as absent in all three expert elicitations (2011, 2013, and 2015). Moreover, the previous panels identified several issues that were still significant matters of concern in 2015, including corruption, arbitrary personalistic rule (persistence of subnational rule by strong power brokers, patronage networks, and capricious decisions by those with authority), and the perverse incentives of elites for maintaining the conflict, as well as Afghanistan's dependence on external supporters.

At the time of this writing, COIN efforts in Afghanistan continued against a backdrop of numerous transitions, with international forces reducing their footprint and numbers and more responsibility being transferred to indigenous institutions and forces. Afghan security forces are now unambiguously the primary COIN forces, where, previously, international forces have had (or shared) that role. The drawdown of international forces was accompanied by an overall reduction in international aid and support, leading to a significant economic transition as well. Perhaps more importantly, these changes were taking place during a time of political transition in Afghanistan, as the Karzai regime gave way to an Afghan government led by Ashraf Ghani and supported by his erstwhile political opponent, Abdullah Abdullah.[14] On the insurgent side, the acknowledgment of Mullah Omar's death sowed confusion and division within the Taliban, and it was no longer clear who led the group and, thus, who claimed responsibility for major decisions, such as entering peace negotiations.[15]

In 2015, Pakistan continued to offer sanctuary to an array of terrorist and insurgent organizations, directly enabling these groups to replenish much-needed resources (though to be sure, there had been a subtle shift in Pakistan in the spring of 2015). The experts judged that

---

[12] Paul, 2011.

[13] Christopher Paul, Colin P. Clarke, Beth Grill, and Molly Dunigan, *Counterinsurgency Scorecard: Afghanistan in Early 2013 Relative to Insurgencies Since World War II*, Santa Monica, Calif.: RAND Corporation, RR-396-OSD, 2013a. It should be noted that while the Delphi exercises over the three years of this study may have had some members in common, they were principally different expert panels and thus may have had slightly different views.

[14] Jonah Blank, "Give Ghani a Chance: Why This Time is Different," *Foreign Affairs*, March 31, 2015.

[15] Joseph Goldstein and Taimoor Shah, "Death of Mullah Omar Exposes Divisions Within Taliban," *New York Times*, July 30, 2015. See also Max Abrahms, "Mohammad Omar's Death Could Help the Afghan Peace Process—or Harm It," *Washington Post, Monkey Cage* blog, August 7, 2015.

governance in Afghanistan had improved ever so slightly, due in large part to Karzai's exodus and his replacement's pedigree as a Western-educated technocrat who knows how to operate in Afghanistan's complex tribal society.

2015 Scorecard Results

The 2015 expert elicitation provided scores for the 42 specific factors in the scorecard necessary to calculate the 15 good and 11 bad factors and practices listed in Table 1.[16] Tables A.2 and A.3 in the appendix show all 42 subfactors and how they roll up into the 15 good and 11 bad factors, respectively. Where there was consensus on the presence or absence of a factor, it was scored "1" if present or "0" if absent; where there was not consensus among the panelists, factors were scored "0.5" for neither present nor absent.[17]

The Delphi panel's scorecard indicates that 7.5 good scorecard factors and 5.5 bad factors were present in Afghanistan in early 2015. The 7.5 good factors include three half-points and thus represent a range from 6 to 9. The 5.5 bad factors include one half-point, thus representing a range between 5 and 6. Subtracting 5.5 from 7.5 provides a scorecard net result of +2. (With all the half-points maximized in one direction or the other, this represents a range from 0 to +4, a band of uncertainty that can be interpreted as something similar to a confidence interval.)

Table 3 shows where these scores (7.5 good, 5.5 bad, +2 net) would fit in among the 59 core insurgencies from *Paths to Victory*. Table 3 lists a subset of the cases in Table 2; it includes only the cases with similar good, bad, and net scores to those found for early 2015 Afghanistan. (See Table 2 for the full list of 59 core cases.) An overall scorecard score of +2 is tied with the lowest historical score for a case won by the COIN force. Having 7.5 good factors puts early 2015 Afghanistan in the middle of pack of COIN winners, several points above the floor for COIN wins (4). Even accounting for uncertainty due to half-points, the lower bounds for the good-factor estimate (6) is consonant with all historical COIN wins. Having 5.5 bad factors, however, puts early 2015 Afghanistan 1.5 points below the record of the worst-scoring historical COIN winner (Guatemala, at –4) and is much more typical of scores for cases in which governments lost. (Four bad factors is the most optimistic bound for the Delphi estimate for contemporary Afghanistan, discounting half-points.)

Without taking into account specific factors or other findings from the broader research, the scorecard results could be interpreted as a source of optimism. The number of good factors present is strong; even considering the lower bound of the uncertainty band due to nonconsensus half-points (a range of 6 to 9), only historical winners had at least six good factors. The total score (good minus bad), +2, is above the historical threshold for a COIN win and is tied with the lowest historical score. The overall band of uncertainty (0 to 4) does dip below the lowest-

[16] Note that the expert elicitation considered 42 factors rather than 15 + 11 = 26 factors because some are summary factors (such as "COIN force realizes at least two strategic communication factors") that rely on multiple subordinate factors for calculation.

[17] Consensus, for this exercise, was considered agreement among at least 70 percent of panelists. In practice, for most consensus factors, degree of accord was considerably higher. The raw averages (equivalent to the percentage of panelists in agreement) are listed in Table A.1 in the appendix.

Table 3
Relevant Subset of Cases and Scores with Early 2015 Afghanistan Scores Overlaid

Country (Insurgency)	Years	Good COIN Practices	Bad COIN Practices	Total Score	Outcome
Nicaragua (Somoza)	1978–1979	0	–6	–6	COIN loss
Chechnya I	1994–1996	1	–7	–6	COIN loss
Bosnia	1992–1995	0	–6	–6	COIN loss
Laos	1959–1975	2	–7	–5	COIN loss
Nagorno-Karabakh	1992–1994	0	–5	–5	COIN loss
Democratic Republic of the Congo (anti-Kabila)	1998–2003	1	–5	–4	COIN loss
Rwanda	1990–1994	2	–6	–4	COIN loss
Bangladesh	1971	2	–6	–4	COIN loss
Afghanistan (Taliban)	1996–2001	2	–6	–4	COIN loss
Kampuchea	1978–1992	0	–3	–3	COIN loss
Cuba	1956–1959	3	–6	–3	COIN loss
Eritrea	1961–1991	1	–4	–3	COIN loss
Sudan (SPLA)	1984–2004	1	–4	–3	COIN loss
Afghanistan (anti-Soviet)	1978–1992	2	–5	–3	COIN loss
Burundi	1993–2003	1	–3	–2	COIN loss
Yemen	1962–1970	1	–3	–2	COIN loss
Lebanese Civil War	1975–1990	5	–7	–2	COIN loss
Tajikistan	1992–1997	2	–4	–2	COIN loss
Nepal	1997–2006	3	–4	–1	COIN loss
Indonesia (East Timor)	1975–2000	3	–4	–1	COIN loss
Nicaragua (Contras)	1981–1990	3	–4	–1	COIN loss
Papua New Guinea	1988–1998	2	–3	–1	COIN loss
Afghanistan (International Security Assistance Force [ISAF])	2001–present	7.5	–5.5	2	***Ongoing***
Iraqi Kurdistan	1961–1975	4	–2	2	**COIN win**
Western Sahara	1975–1991	4	–2	2	**COIN win**
Argentina	1968–1979	5	–2	3	**COIN win**
Oman (Imamate Uprising)	1957–1959	4	–1	3	**COIN win**
Croatia	1992–1995	5	–2	3	**COIN win**
Guatemala	1960–1996	8	–4	4	**COIN win**
Tibet	1956–1974	7	–3	4	**COIN win**
Sri Lanka	1976–2009	6	–1	5	**COIN win**
Mozambique (RENAMO)	1976–1995	8	–3	5	**COIN win**
Turkey (PKK)	1984–1999	8	–2	6	**COIN win**
Indonesia (Aceh)	1976–2005	8	–2	6	**COIN win**

NOTE: Shading denotes scores comparable with Afghanistan's in 2015, with green shading indicating cases with comparable numbers of good factors, red shading indicating comparable numbers of bad factors, and blue indicating a comparable overall score.

scoring historical winners (+2 being the lowest winner's score), but it does not quite touch the highest-scoring historical losers (–1 being the highest historical loser's score).

However, even the summary results from the scorecard offer cause for concern. The fact that the sum of bad factors (5.5) is greater than the sum of bad factors for any historical winners is troubling. Also concerning is the fact that Afghanistan has a relatively large number of factors present from both lists (7.5 good + 5.5 bad = 13 total factors present). Other cases with a high number of total scorecard factors had a preponderance on one side or the other. Other cases with relatively balanced scorecards had relatively modest totals on each side. The Afghanistan scorecard is in much greater tension than the simple summary "overall +2" implies.

Fortunately, the scorecard and elicitation exercise offer additional value in identifying the specific factors that are present or absent and some of the implications thereof. While the raw scorecard score might suggest optimism about prospects for success in Afghanistan, our overall analysis is much more pessimistic. The absence of certain critical factors (and the lack of any real prospects for improving on those factors) leads us to conclude that a military victory by Afghan COIN forces is unlikely. A mixed outcome remains much more likely, with some kind of negotiated settlement being the quickest path to conflict resolution—a path that could lead to a resolution on terms favorable to the government. The next section discusses these factors and provides suggestions regarding which factors might be improved upon. By way of preview, critical factors that are absent include a demonstration of commitment and motivation on the part of the Afghan government and Afghan security forces and the interdiction of tangible support to the insurgents (interrupting flows of materiel and other forms of support both from within Afghanistan and across borders).

Specific Factors in the Current Case

Tables A.2 and A.3 in the appendix present all factors and subfactors in the scorecard and the factors scored as present or absent during the expert elicitation exercise. In each row, a "1" indicates that the consensus view was that the factor or practice was deemed to be present in the current conflict; a "0" indicates that that the consensus view was that the factor was absent.[18] When reviewing the scores, remember that participants were asked to make worst-case assessments; isolated anecdotes or limited accounts of success in these areas were not considered sufficient grounds to score a factor (be it a bad factor or a good factor) as present. Where the score is "0.5," there was no clear consensus among participants. Consensus was reached for 33 of the 42 factors scored directly by the panel. These factors are discussed in greater detail later. Rows beginning with numbers are primary factors (i.e., the top-level factors listed Table 1); rows beginning with lowercase letters are the subfactors that constitute the primary factors. (All subfactors were directly scored by panelists, as were any primary factors without subfactors.)

The discussion and findings from our foundational historical research suggest that some of the good factors that the participants identified absent are particularly concerning.

Early 2015 Afghanistan was scored as having 7.5 of a possible 15 good primary factors present and, thus, 7.5 (actually, six and three half-points) absent. The absent factors and those for which there was not consensus agreement of presence or absence are as follows (with those for which there was not consensus denoted with a 0.5 in parentheses):

---

[18] In all cases, the threshold for consensus was 70-percent agreement among the expert participants. The percentages of participants indicating each factor as present or absent are listed in Table A.1 in the appendix.

- COIN force realizes at least two strategic communication factors (from a list of five strategic communication factors: COIN force and government actions consistent with messages; COIN force maintains credibility with populations in the area of conflict; messages/themes coherent with overall COIN approach; COIN force avoids creating unattainable expectations; themes and messages coordinated for all involved government agencies).
- COIN force reduces at least three tangible support factors (from a list of ten tangible support factors: flow of cross-border insurgent support significantly decreased, remains dramatically reduced, or largely absent; important external support to insurgents significantly reduced; important internal support to insurgents significantly reduced; insurgents' ability to replenish resources significantly diminished; Insurgents unable to maintain or grow force size; COIN force efforts resulting in increased costs for insurgent processes; COIN forces effectively disrupt insurgent recruiting; COIN forces effectively disrupt insurgent materiel acquisition; COIN forces effectively disrupt insurgent intelligence; COIN forces effectively disrupt insurgent financing).
- Government realizes at least one government legitimacy factor (from these two: government leaders selected in a manner considered just and fair by majority of population in the area of conflict; majority of citizens in the area of conflict view government as legitimate).
- Government corruption reduced/good governance increased since onset of conflict.
- Unity of effort/unity of command maintained (0.5).
- COIN force avoids excessive collateral damage, disproportionate use of force, or other illegitimate applications of force (0.5).
- COIN force established and then expanded secure areas.
- COIN force provides or ensures provision of basic services in areas it controls or claims to control.
- Perception of security created or maintained among population in areas COIN force claims to control.
- Majority of population in area of conflict supports/favors COIN forces (0.5).

These absences suggest possible areas for improvement. However, some of these absent factors are more concerning than others. While the COIN scorecard simply assumes equal weight across all scorecard factors (and successfully discriminates the core historical cases into wins and losses with that assumption in place), other aspects of the *Paths to Victory* research show that, in fact, some of these factors are more important than others. One of these factors, found to be a prerequisite to success in all of the historical cases, was again scored as absent in Afghanistan in 2015: the interdiction of tangible support to the insurgents. This concerns the extent to which COIN forces have cut flows of support (e.g., materiel, personnel, financing, intelligence) to the insurgents. The expert panel scored nine of the ten tangible support reduction subfactors as absent and did not reach consensus on the tenth (COIN force efforts resulting in increased costs for insurgent processes). For the parent factor to be scored "1" on the scorecard, at least three of the subfactors would have had to be scored as present (and, obviously, even more would be better). This is particularly concerning because, in the 59 core historical cases, all the COIN winners disrupted at least three tangible support factors and none of the losers did. The discussion revealed that the insurgents in Afghanistan meet their tangible support needs from many sources, none of which had been significantly disrupted: supporters across the border in Pakistan (and in other countries),[19] profits from drug trafficking and other criminal activities, and supportive populations (sometimes coerced) inside Afghanistan. Disrupting tangible support would require significant efforts to interrupt support flows from all of these sources, but this may well be necessary for COIN forces to prevail.

Along with the absence of certain critical good factors, the presence of several critical bad factors suggests possible areas for improvement. Five (and one half) of the 11 bad factors were scored as present by the expert panel:

- Government involves corrupt and arbitrary personalistic rule.
- Country elites have perverse incentives to continue conflict.
- Host nation is economically dependent on external supporters.
- Fighting primarily initiated by the insurgents.
- Insurgent force individually superior to the COIN force by being either more professional or better motivated (0.5).
- COIN force and government have different goals/level of commitment.

Not only is the total number of bad factors present concerning, but so are the factors themselves. Most of the bad practices are from a single category from the previous research: commitment and motivation. A government and COIN force committed to fighting the insurgency was also present in all historical cases won by the government, and thus another critical prerequisite to COIN success appears to be absent in Afghanistan. The extent to which the government and COIN forces facing the insurgency demonstrated their resolve to defeat that insurgency is one of the strongest predictors of success in the historical cases. In the historical cases, no government with four or more of eight factors indicating low commitment and motivation managed to prevail. The expert panel agreed that early 2015 Afghanistan had too many of these factors present.[20] This suggests that the Afghan insurgency can be defeated only if the Afghan government and security forces are able to establish and demonstrate their commitment to doing so. This is a real concern, as a total collapse of the Afghan National Security Forces (ANSF) or an unwillingness to fight could lead the insurgents to retake strategic provinces or, worse, allow the Islamic State to gain a foothold in parts of eastern Afghanistan along the border with Pakistan.

Notes on Factors Present or Absent in 2015 but Tenuous in the Future

The 2013 Delphi exercise and the discussion during the expert elicitation often highlighted factors that might become precarious in the future as international forces and support continued to draw down. During the 2015 exercise and discussion, the following factors were cited as meriting at least some future concern in the face of ISAF retrograde, regardless of whether one or more panelists asserted the factors were favorably present or absent in Afghanistan in early 2015:

- Themes and messages coordinated for all involved government agencies (panel failed to reach consensus on presence or absence).

---

[19] U.S. Department of Defense, *Progress Toward Security and Stability in Afghanistan*, Washington, D.C., October 2014, p. 7.

[20] Of the eight commitment and motivation factors referenced, only six are included in the scorecard. The expert panel agreed that four of those six were present, and consensus was not reached on one other (half-point).

- Insurgents' ability to replenish resources significantly diminished (absent by consensus).
- COIN force efforts resulting in increased costs for insurgent processes (panel failed to reach consensus on presence or absence).
- COIN forces effectively disrupt insurgent recruiting (absent by consensus).
- Intelligence adequate to support kill/capture or engagements with insurgents on COIN force's terms (panel failed to reach consensus).
- COIN force of sufficient strength to force insurgents to fight as guerrillas (present by consensus).
- Unity of effort/unity of command maintained (panel failed to reach consensus).
- Short-term investments, improvements in infrastructure/development, or property reform in area of conflict controlled or claimed by COIN force (present by consensus).
- COIN force establishes and then expands secure areas (panel failed to reach consensus).
- Government/COIN reconstruction/development efforts sought/achieved improvements substantially above historical baseline (present by consensus).
- COIN force provides or ensures provision of basic services in areas it controls or claims to control (panel failed to reach consensus).
- Insurgent force individually superior to COIN force by being either more professional or better motivated (panel failed to reach consensus).

This finding is concerning because the current scorecard balance of 7.5 good factors, 5.5 bad factors, and an overall score of +2, with attendant bands of uncertainty, indicates a need for further improvement. These concerns cast significant doubt on the prospects for further improvement once ISAF withdraws and, in fact, imply a significant threat of sliding backward toward even less optimistic scores.

In addition to the possibility that Afghan forces cannot sustain the presence of some of these factors, there is also a concern about their desire to do so. Numerous observers have suggested that the Afghans conceive of the insurgency differently from Westerners, and now that they are unambiguously in charge of the COIN effort, that effort may change in line with their different perception of the situation. Choosing to emphasize or focus on different practices could also significantly affect the presence or absence of key factors going forward.

2015 Results Compared with Previous Results and Analyses

As noted, this report repeats two previous efforts to complete the COIN scorecard for Afghanistan; the previous findings were published in *Counterinsurgency Scorecard: Afghanistan in Early 2011 Relative to the Insurgencies of the Past 30 Years,* based on research reported in *Victory Has a Thousand Fathers: Sources of Success in Counterinsurgency,* and *Counterinsurgency Scorecard: Afghanistan in Early 2013 Relative to Insurgencies Since World War II,* based on research reported in *Paths to Victory: Lessons from Modern Insurgencies.*[21]

The COIN scorecard has an observed "empirical gap" separating the highest-scoring win from the lowest-scoring loss—a gap of uncertainty between the outcomes actually observed. In the 2011 Afghanistan scoring, the scorecard result was found to be in that gap between historical wins and losses. Results from the 2011 exercise showed that Afghanistan's good factors

[21] Paul, 2011; Paul, Clarke, and Grill, 2010; Paul, Clarke, Grill, and Dunigan 2013a; Paul, Clarke, Grill, Dunigan, 2013b.

had it tied with Croatia (1992–1996) as the lowest-scoring winner in terms of good factors; it was also close to the worst-scoring winner (Turkey [PKK] [1984–1999]) in terms of bad factors, but also close to many losing COIN forces (DR Congo [anti-Kabila] [1998–2003]) and Nagorno-Karabakh [1992–1994]). In the 2013 scoring, the result crossed that gap, and, at +2, Afghanistan scored among the historical COIN winners. Participants in 2013 conveyed that several important areas had seen improvement since the 2010–2011 time frame. Although Afghanistan's total score in 2013 was tied with the lowest-scoring historical winners, the band of uncertainty (like a confidence interval) due to nonconsensus half-point scores, ranged from –2 to +6, covering both the lowest-scoring historical wins and the highest-scoring historical losses. The 2015 score is also +2, though now with the range of uncertainly narrowed to no longer cover any historical losses. It does still extend into the gap of uncertainty between the highest-scoring historical loser and the lowest scoring historical winner, however.

The 2011 report highlighted three critical areas for improvement: (1) the competency, legitimacy, and popular support of the Afghan government; (2) security; and (3) disrupting tangible support to the insurgents. The 2015 exercise reiterated two critical areas for improvement that were also stressed in the 2013 results, one of which matches the 2011 study: (1) disrupting tangible support to the insurgents and (2) demonstration of commitment and motivation on the part of the Afghan government and security forces. Compared with the 2011 panelists, both the 2013 and 2015 panelists were much more cognizant of the impending departure of the majority of coalition forces and the corresponding challenges to prospects for maintaining the good practices that were in place at the time of the earliest study.

Between 2013 and 2015, Afghanistan's overall scorecard score remained +2, but there were several interesting changes at the factor level that merit discussion. From a security force perspective, several factors improved, several declined, and two contentious factors remained contentious. Perhaps most importantly, COIN forces were still unable to disrupt the tangible support of the insurgents as of early 2015, and participants discerned only slight improvement in the commitment and motivation of the government and the COIN force. Still, improvement in other areas, including consistency of messages, increased legitimacy, and improved unity of effort could very well have spillover benefits for the commitment and motivation of the force if these gains are real and can be properly channeled. None of these changes (positive or negative) are a *fait accompli*, and as demonstrated by the factors that declined and the negative factors that remained steady without improvement, gains are hard won and easily reversed.

What Improved?

In 2015, the expert panel found that several key factors improved from just two years earlier, though it should be stressed that each of these improvements was marginal, increasing from absent to contentious in the case of good factors and from contentious to absent in the case of one bad factor. The panel noted improvements from absent to contentious in the following factors: "COIN forces and government actions consistent with messages (delivering on promises)," "Unity of effort/unity of command of COIN forces maintained," "Government leaders selected in a manner considered just and fair by majority of population in the area of conflict," and "Majority of citizens view government as legitimate in the area of conflict." The latter two factors likely benefited from the "Ghani bump," an infusion of optimism that may have owed as much to the fact that Ashraf Ghani is not Hamid Karzai as to the public's confidence in Ghani's ability to lead Afghanistan out of its current morass. The one bad factor deemed to no longer be a significant issue was "COIN forces or allies rely on looting for sustainment," which could hint at a growing professionalism among the ranks of the ANSF, or it could simply be wishful thinking on the part of some of the panelists.

What Declined?

One of the most significant changes across the three scorecards was the transition in who was the primary COIN force. In 2011, ISAF forces had the lead in most operations; by 2013, ISAF forces partnered to a great extent with the ANSF, but international forces still shouldered most of the burden. In 2015, there were still some international forces in country, but the mantle of primary counterinsurgent is now unambiguously on the shoulders of the ANSF. Some of the positive factors showing declines on the 2015 scorecard are undoubtedly related to the military competence of the ANSF relative to ISAF forces.

As with the improvements discussed earlier, the factors that declined also did so at an incremental rate, moving from scores of present to contentious for factors that, if scored present, would have been positive indications for the Afghan government and COIN forces. The three factors that changed sufficiently to move from present to contentious were "COIN force efforts resulting in increased costs for insurgent processes," "COIN force avoids excessive collateral damage, disproportionate use of force, or other illegitimate applications of force," and "Majority of population in area of conflict supports/favors COIN forces over the insurgents." A major concern is that changes in these factors, if reflective of the situation on the ground, seem to indicate an ongoing stalemate between the ANSF and the insurgents.[22] And while a stalemate is worrisome from the perspective of achieving a military victory over the insurgents, it also offers a potential opportunity to engage the insurgents in talks and reach a negotiated settlement to the conflict.[23] But if U.S. and coalition assistance, in the form of military training and advising and foreign aid, cannot be sustained at current levels for the foreseeable future, critical factors may continue to decline and thus affect the ongoing stalemate. If the Taliban perceive that they are gaining leverage, the chances for brokering a lasting negotiated settlement will degrade.

What Stayed the Same?

Two factors that were scored as contentious in 2013 remained contentious during this iteration. Those factors were "Messages/themes coherent with overall security approach" and "Insurgents individually superior to the COIN forces by being either more professional or better motivated," both of which held steady at 0.5. That certain factors remained contentious in 2015 is unsurprising. Indeed, Afghanistan remains a complex patchwork of progress, stagnation and regression. The side that maintains the "upper hand" in a certain province can vary from week to week, month to month. For those seeking to understand what these factors portend, it is critical to broaden the aperture of analysis and probe the deeper meanings and fundamental reasons for some of these changes, or lack thereof.

What are some possible explanations for the improved standing of Afghanistan's government and security forces? Some experts argued that years of consistency in areas like training, equipping, and funding are finally paying dividends, noting the inherent lag time for these

---

[22] Jim Michaels, "Decade of War, Billions in U.S. Aid Fail to Defeat Taliban," *USA Today*, May 19, 2015.

[23] Clarke and Paul, 2014.

effects to manifest with respect to discipline and battlefield performance.[24] Skeptics might counter that any perceived improvements are merely the result of continued oversight by the United States, even though U.S. troops are mostly relegated to a counterterrorism role—tracking down al Qaeda militants and training and advising the Afghan units responsible for the lion's share of the fighting throughout the country.[25]

Scorecard and Duration of Conflict

The transition from within the uncertainty gap to among the historical winners between 2011 and 2013 is very interesting in light of other findings concerning the scorecard from *Paths to Victory*. When we briefed results from the original scorecard (detailed in *Victory Has a Thousand Fathers*), we would often be asked something like, "How long does the COIN force need to keep a strong scorecard balance before it wins?" We were unable to answer this question with our initial data, but we could do so with the new data collected for the follow-on study (*Paths to Victory*). The answer is, on average (median), 5.75 years. That is, in the 28 insurgencies won by governments since World War II, all had scorecard balances of +2 or greater by the end of the conflict, and, on average, the conflict ended 5.75 years after achieving a scorecard balance of +2 or greater. There was considerable variation in this average, with some cases having a positive scorecard balance for only a short time before a decisive outcome and others having a positive scorecard balance for the entirety of a conflict running 12 years or more. But the average was 5.75 years.

This has important implications for Afghanistan. If we impute the 2011 results as falling short of the needed threshold and consider the 2013 results as finally reaching a minimally sufficient scorecard balance, then history suggests that a positive scorecard balance in 2015 would need to be sustained for an additional several years before the conflict's conclusion favors the government (3.75 more years, so until at least late 2018, but perhaps much longer). The expert panelists' concerns about the ability of the Afghan government and security forces to sustain some of the good COIN practices after the departure of the majority of international forces call into question whether the Afghans can maintain a positive scorecard balance by themselves and, ultimately, achieve a favorable outcome.

Summary of Scorecard Analyses

What do the scorecard results mean for the way ahead in Afghanistan? Previous RAND studies have made the general recommendation that COIN forces should always seek to maximize good factors and minimize bad factors until the conflict is resolved. In 2011, scorecard-based recommendations for the conflict in Afghanistan focused on increasing the legitimacy, compe-

---

[24] One of the key findings from recent RAND research on building partner capacity noted that "consistency is key—not just consistent funding, but consistent objectives, consistent agreements, and consistent relationships" (Christopher Paul, Jennifer D. P. Moroney, Beth Grill, Colin P. Clarke, Lisa Saum-Manning, Heather Peterson, and Brian Gordon, *What Works Best When Building Partner Capacity in Challenging Contexts?* RR-397-OSD, Santa Monica, Calif.: RAND Corporation, 2015, p. xii).

[25] Azam Ahmed and Joseph Goldstein, "Taliban Gains Pull U.S. Units Back into Fight in Afghanistan," *New York Times*, April 29, 2015.

tency, and popular support of the Afghan government (including service provision); improving security (successfully and consistently establishing and expanding secure areas and creating and maintaining a perception of security throughout the area of conflict); and disrupting the flow of tangible support to the insurgents (including recruits, materiel, financing, and intelligence). The recommendations were similar in 2013, highlighting the continued need to disrupt the tangible support of the insurgents (both within and across borders) while also concentrating on the commitment and motivation of the Afghan government and security forces.

In 2015, the scorecard suggested some improvement and progress, but also many areas for concern. The experts found that prospects for sufficiently reducing the insurgents' tangible support to bring about an end to the conflict were low, so achieving an unambiguous military victory was unlikely. The best the government of Afghanistan can likely hope for is a favorable negotiated settlement—a mixed outcome favoring the government.

Prospects for a Negotiated Settlement

Our research on resolving insurgencies through negotiated settlements is relevant to 2015 Afghanistan. Surveying 13 cases of conflict resolution, we sought to identify a master narrative that captured the essential ingredients and sequence illustrated in most of them; we then applied the master narrative approach to Afghanistan.

As explained earlier and illustrated in Figure 1, the master narrative for negotiated settlements unfolds in seven steps generally, though not always in this exact order. First, after years of fighting, both sides to the conflict reach a state of war weariness and settle into a mutually hurting military stalemate. Second, after a stalemate has been reached and the belligerents recognize the futility of continued escalation, the insurgents are accepted as a legitimate negotiating partner. Once the insurgents have been accepted by the government, the terms of a ceasefire can be discussed. This third step, ceasefire, like step 2 before it, is highly dependent on the acquiescence of external powers. For example, if an active external supporter is pushing for continued conflict, it is likely that the negotiation process will end here.

If external actors refrain from further meddling, official intermediate agreements can be reached, the fourth step in the narrative. Fifth, power-sharing offers (such as amnesty or elections) can further entice the insurgents to favor politics over armed struggle. Sixth, once offers of power-sharing have been accepted, a moderation of the insurgency's leadership can facilitate further progress by giving a voice to the politically minded cadre of the group. Seventh, and finally, third-party guarantors are required to help guide the process to a close, acting as impartial observers or providers of security, economic assistance, development aid, and other services.

A detailed description of the process and case studies that led to the development of the master narrative can be found in the report *From Stalemate to Settlement: Lessons for Afghanistan from Historical Insurgencies That Have Been Resolved Through Negotiations*.[26] While only one of the cases considered unfolded exactly according to this sequence, each case unfolded in a manner close enough to this narrative that it is a useful comparative tool for understanding how to get to negotiated settlements.

---

[26] Clarke and Paul, 2014.

Given the conclusion that a negotiated settlement may be the most promising route to ending the conflict in Afghanistan, it is most useful to compare Afghanistan's scorecard to those of other cases in which mixed outcomes led to negotiated settlements. As of early 2015, Afghanistan was unquestionably heading toward a "mixed" outcome, and the prevailing view among U.S. policymakers is that the best and only outcome is a political settlement. The following cases hold potential lessons for Afghanistan:

- *Western Sahara:* Like Afghanistan, Western Sahara had an overall score of +2. One of the critical factors in achieving a military stalemate was improved border security through a system of defensive sand walls—clearly not a realistic option in the Afghan case. One interesting parallel between the two cases, however, was that an agreement by Moroccan forces to withdraw from disputed territory created the momentum that led to talk of a referendum and gave the warring parties an opportunity to map out and eventually implement what became known as "the Settlement Plan" over the three years following the withdrawal of troops.
- *Tajikistan:* This case finished with an overall score of –2, so four clicks in the other direction from Afghanistan, with 2015 Afghanistan reflecting five bad practices, compared with four for the Tajik COIN force. More importantly, this case highlights the importance of an external actor. Once Russia decided that a conflict simmering on its border was too risky, Moscow intervened forcefully in favor of a political settlement. In Afghanistan, the primary external actor with a similar level of clout is Pakistan, though, in this case, the external actor would be pressuring the insurgents and not the government to reach a deal.
- *Lebanese Civil War:* Similar to Tajikistan, the overall score of –2 put the mixed-outcome Lebanese case a mere four points away from 2015 Afghanistan. Most instructive in the Lebanese case is that although the civil war ended in 1990, one of the major protagonists—Hezbollah—remains a powerful politico-military force within the country, essentially a state within a state in southern Beirut and its surrounding environs, never making a true transition to a fully integrated part of the Lebanese state. Some speculate that Israel's 18-year occupation of Lebanon contributed to Hezbollah's legitimacy and staying power, as resisting the Israeli became the Shia militia's *raison d'être.*
- *Yemen:* With an overall score of –2 and a mixed outcome (though favoring the insurgents), the Yemeni case provides another clear example of troop withdrawal used as a bargaining chip to move from stalemate and to political settlement. In Afghanistan, it is worth strongly considering a policy of "deliberate and incremental de-escalation," whereby foreign troop levels would be subject to negotiation as a significant bargaining chip. It follows that the Taliban could achieve a gradual and guaranteed reduction by participating in a process of talks and eventual ceasefires, demonstrating measurable progress and providing momentum for a political settlement through a serious of tangible confidence-building measures.
- *Burundi:* Still another case with an overall score of –2 with a mixed outcome that resulted in a negotiated settlement, Burundi offers a critical but difficult-to-achieve example of quid pro quo. In that case, the insurgents consented to the demobilization of its fighters in return for ministry positions, diplomatic posts, and local government control as part of an agreed-upon power-sharing deal. If a similar program could gain traction in Afghanistan,

this would greatly enhance prospects for a sustainable solution, though long-term peace and stability are never a guarantee.

As of late May 2015, there were reports of preliminary meetings between Afghan government officials and former Taliban officials with close ties to Pakistan's Inter-Services Intelligence bureau in Urumqi, in China's Xinjiang province.[27] Whether these talks prove fruitful, the prospects for a negotiated settlement will remain reasonable if COIN forces are able to continue fighting effectively enough to achieve at least a mutually hurting stalemate—a delicate balance tested each fighting season in Afghanistan. At this stage in the conflict, outright military victory over the insurgents is neither a realistic possibility nor a stated objective of Afghan or ISAF policy. Indeed, the best military outcome will be an ongoing stalemate that offers an opportunity to engage the insurgents in exploring the factors and variables necessary to reach a negotiated settlement to the conflict.[28] That is what success looked like by late 2015, with the transition to the North Atlantic Treaty Organization's (NATO's) Resolute Support Mission in Afghanistan, Operation Freedom's Sentinel (which replaced Operation Enduring Freedom at the end of 2014), and the transition from coalition military to Afghan civilian control.

Ambassador James Dobbins and Carter Malkasian suggest several critical concrete steps to keep negotiations in play, the most important of which is a compromising concession. They suggest that the Afghan government needs to come to terms with the fact that the Taliban will be a legitimate entity in a new government that might include constitutional reform or new institutional arrangements.[29] The acknowledgment in the summer of 2015 that Mullah Omar had died back in 2013 could certainly complicate—or, at the very least, delay—negotiations with the Taliban.[30] Rather than seizing upon this recent development as a point of leverage, Afghanistan's government should continue to propose other steps that will be perceived as a true compromise, not simply enhancing the government's position vis-à-vis the Taliban. Without more aggressive steps to promote a settlement, the military dynamic could slip in the Taliban's favor, and the prospects of a durable settlement would, in turn, fade.

In the meantime, a sustained commitment from coalition forces and the international community can demonstrate resolve, but there must be a transparent discussion about the role and level of external forces in Afghanistan going forward. Instead of following a policy of strategic ambiguity with respect to when military forces might be withdrawn and exactly how many will remain in country, the United States should pursue deliberate and incremental de-escalation whereby the foreign troop level would be "on the table" and the Taliban could achieve a gradual and guaranteed reduction by participating in talks and eventual ceasefires. The issue of external forces is a valuable bargaining chip and a huge motivating factor for the insurgents to remain true to a potential settlement. This process is not without complications, however. It is not plausible for the United States to push the Taliban to the table by declaring that it will leave behind a residual force of 10,000 troops in Afghanistan. Furthermore, there is an obvious tension between putting troop withdrawal on the table as a bargaining chip and

---

[27] Edward Wong and Mujib Mashal, "Taliban and Afghan Peace Officials Have Secret Talks in China," *New York Times*, May 25, 2015.

[28] Clarke and Paul, 2014.

[29] James Dobbins and Carter Malkasian, "Time to Negotiate in Afghanistan," *Foreign Affairs*, July–August 2015.

[30] Anthony H. Cordesman, *The Afghan Campaign and the Death of Mullah Omar*, Washington, D.C.: Center for Strategic and International Studies, August 2, 2015.

the continued presence of Islamic State fighters, with the real concern that withdrawing troops in earnest could lead to a power vacuum similar to 2011 Iraq.

In an ideal setting, putting foreign troop reduction on the table would allow the insurgents to commit to some version of a narrative that allows its leaders to save face while making gradual steps toward winding down the fighting, probably in return for some version of a power-sharing role or political stake in a future Afghanistan. A successful and enduring negotiated settlement almost certainly requires not only the support of the core actors (e.g., the Afghan government, Taliban, Pakistan, United States) but also buy-in from regional powers, including Iran, Russia, and India (what Shinn and Dobbins call first-ring actors).[31] Chinese participation also seems critical at this point in the process.

Finally, after taking stock of the good and bad scorecard factors, it appears that there is a need to invest in policy efforts that can positively affect a negotiated political settlement, focusing on government legitimacy factors, government corruption (reducing the incentives of elites on all sides to continue the conflict), and strategic communication factors. The latter are even more critical during the long and arduous process of crafting a negotiated settlement, since achieving a settlement is often a precarious process, characterized by myriad opportunities for misunderstanding and accusations of duplicity. As of this writing, it was still too early to tell whether the conflict in Afghanistan would end through a negotiated settlement. Should this prove to be the case, it seems likely that the result of any negotiated settlement will be "mixed," with both the Afghan government and the Taliban making fairly major concessions to reach an agreement. Even if this is not the ultimate endgame, it is still valuable to examine the right combination of factors needed for a negotiated settlement. While not a *fait accompli*, it is becoming increasingly apparent that Afghanistan is more likely to end in a negotiated settlement than outright victory by either side. The situation as it stands today is around step 2 in the process, which means that some combination of the approximately five remaining steps toward a negotiated settlement still need to occur.

To reach a negotiated settlement, it helps if both sides have faced setbacks, neither side perceives unambiguous military victory as likely, external actors reduce support to both sides, and all external actors press for a negotiated settlement (and at least one is willing to act as a guarantor). Based on these criteria, prospects for a negotiated settlement in Afghanistan look reasonably good if the insurgents are unable to prevail militarily once the coalition withdraws, if insurgents' external supporters push for a negotiated settlement, and if a third party can be found to act as an honest broker and provide peacekeepers.

With a stalemate largely achieved, the "master narrative" of reaching a negotiated settlement as applied to Afghanistan suggests the following efforts are necessary to make progress toward a negotiated settlement (steps 2–5):

- *Step 2:* Leading external powers co-opt the leadership on both sides of the conflict and apply sufficient pressure to accept each other as legitimate negotiating parties (the United States with the Ghani administration, and Pakistan with the Taliban)
  - *Substep 2:* The COIN force and its external partners convince other external powers to support peace rather than continued fighting.
- *Step 3:* The two sides or designated external parties broker a ceasefire.

[31] James Shinn and James Dobbins, *Afghan Peace Talks: A Primer*, Santa Monica, Calif.: RAND Corporation, MG-1131-RC, 2011, pp. 52–59.

- *Step 4:* The two sides make progress toward some type of official agreement.
- *Step 5:* The process ensures that there is the promise of political legitimacy for the insurgent leadership, and not a measure of last resort.

The first half of 2015 witnessed unofficial meetings between the Taliban and the Afghan government in China in May, an official meeting in Pakistan in early July, and yet another meeting was scheduled for late July, just around the time it was announced that Mullah Omar had died, at which point peace talks were indefinitely postponed.[32] It remains unclear exactly what position the new Taliban leader Akhtar Mohammad Mansour will take on a negotiated settlement (though it is not a promising sign that Tayyab Agha, head of the Taliban's political wing, resigned). What is certain is that making progress along the path toward steps 2–5 will require active participation from all parties involved, especially the United States, the Afghan government, China, and Pakistan.

The United States has invested billions of dollars in Afghanistan, in the form of aid for the Afghan government and to support the kinetic part of the fight—countering the insurgency and training and equipping the ANSF. Now, it must channel its influence toward regional diplomacy (while not eschewing the resources necessary to maintain a military stalemate) aimed at bringing the Taliban to the negotiating table. The rise of the Islamic State in Afghanistan, the death of Mullah Omar, and subsequent Taliban infighting have shifted the political dynamic. Establishing confidence-building measures would go a long way to building momentum toward steps 2–5. This could include reopening the Taliban's political office in Qatar, prisoner exchanges, the easing of sanctions, and the guarantee of safe passage for Taliban figures engaged in diplomacy. The ultimate goal of U.S. engagement in engineering a political settlement should be separating the Taliban from al Qaeda and including the Taliban in reconciliation efforts aimed at ending the internally directed insurgency.[33]

Conclusions and Recommendations

Similar to the Afghanistan scorecards of 2011 and 2013, the 2015 scorecard compares present-day Afghanistan with historical cases of insurgency since World War II. This comparison offers some support for optimism, but it also raises some significant concerns. When using a scorecard of 15 good factors and 11 bad factors based on the historical record, Afghanistan's current balance of +2 places it among the historical winners. Further, Afghanistan's current score of 7.5 of 15 positive factors is strong relative to other historical winners.

However, the band of uncertainty around the total score includes both historical wins and historical losses, and the current score of 5.5 bad factors is larger than the number of negative factors possessed by any of the winners. Furthermore, although we can currently count 7.5 good factors as present, two factors that were present in all historical wins are still missing, just as in 2013: the disruption of insurgents' tangible support and a clear demonstration of commitment and motivation on the part of the host-nation government and indigenous security forces. Two final concerns raised by the experts who participated in the 2015 scoring were that the Afghan government and security forces may not be able to maintain several of the

---

[32] Carter Malkasian, "Is Peace Possible in Afghanistan?" *Foreign Affairs*, August 18, 2015.

[33] Eggers, 2015.

good practices once the coalition of international supporters withdraws the preponderance of their forces and that several good COIN practices appear to be present only in certain (densely populated or uncontested) regions of the country.

Finally, there are two relatively new wild cards to consider, including the promise of the Ghani administration and the potential for the Islamic State to gain traction in the country. One major challenge for the Ghani administration will be containing the insurgency to ensure that it is not able to spread beyond traditional areas of Taliban control, something the ANSF was struggling to do in early 2015.[34] Regarding the Islamic State, reports indicate that the group's militants are flocking to Afghanistan to fight coalition forces.[35] At this point, it seems clear that the Taliban will not unite with the Islamic State and that Taliban leadership perceives the group as a threat, but the recent declaration of Mullah Omar's death could cause the Taliban to splinter, opening avenues for the terrorist group's entry into certain parts of Afghanistan.

One possible benefit of a growing Islamic State presence could be a warming of relations between the governments in Kabul and Islamabad, which would have a shared interest in preventing the group from gaining a foothold in the region.[36] This rapprochement was initially spurred by the Peshawar school massacre in December 2014, after which the Pakistani government reevaluated its stance toward certain militant groups operating on its soil. U.S. diplomats should pursue this potential rapprochement with vigor.

The 2015 expert elicitation exercise highlighted two critical shortfalls in the Afghanistan COIN effort: (1) a failure to disrupt flows of tangible support to the insurgents and (2) a failure to demonstrate sufficient commitment and motivation on the part of the Afghan government and Afghan security forces. These shortfalls have been concerns since the onset of the conflict and temper any optimism based on the other scorecard results.

Rather than continuing to recommend that U.S. efforts seek to help Afghanistan address these shortfalls, we offer more nuanced recommendations. The United States should both seek to bolster Afghan COIN efforts *and* take steps to improve the prospects for a negotiated settlement. Fortunately, there is potentially substantial overlap between the two. If Afghan security forces are more successful in their COIN efforts, they will be more likely to maintain a stalemate (the critical first step in the typical sequence for negotiated settlements) and perhaps allow the Afghan government to negotiate from a position of strength. U.S. efforts to pressure the insurgents' external supporters should focus on reducing tangible support, of course (which would improve the prospects of both the COIN effort and a stalemate en route to negotiations). However, they should also focus also on encouraging external actors to support—or at least not actively resist—negotiations (important in many of the steps to settlement).[37]

In a world of finite resources and declining domestic appetite for international commitments, the bulk of resources should be dedicated to bolstering prospects for successful negotiations. As the clock runs out on the Obama administration, it is time to begin thinking about

---

[34] Tim Craig, "Afghan Forces Straining to Keep the Expanding Taliban at Bay," *Washington Post*, May 16, 2015; see also Sudarasan Raghavan, "As the U.S. Mission Winds Down, Afghan Insurgency Grows More Complex," *Washington Post*, February 13, 2015.

[35] Sudarasan Raghavan, "Foreign Fighters Loyal to ISIS are Now Flowing into Afghanistan," *Washington Post*, April 14, 2015.

[36] Katharine Houreld, "Pakistan Military Says Its Spies Will Cooperate with Afghanistan," Reuters, May 19, 2015.

[37] Clarke and Paul, 2014.

the future level of U.S. presence in Afghanistan. In addition to the obvious imperatives to accelerate the conflict's termination while also preventing the collapse of the Afghan government and the reemergence of the country as a hub for transnational terrorists, a critical piece of sustaining even a small footprint in Afghanistan will be retaining domestic support among the U.S. population.[38] Provided force commitments remain modest and U.S. casualties remain low, we anticipate minimal domestic opposition to a continued presence. The situation in Afghanistan is more politically palatable because there is no issue with a status-of-forces agreement, as there was in Iraq, and there is little concern about a U.S. adversary exerting a negative influence on the government in Kabul (again, as there was and continues to be in Iraq with Iranian meddling). But the comparison goes further. To avoid repeating missteps of the conflict in Iraq, the United States should push for reconciliation to ensure an inclusive negotiated settlement (likely to include constitutional reform) that gives the Taliban a legitimate voice in the political process. By achieving this objective, U.S. policy would work to drive a wedge between the internally directed insurgency of the Taliban and the transnational threat posed by al Qaeda, the Islamic State, and other groups that might seek to take advantage of Afghanistan's instability.[39]

---

[38] Stephen Watts and Sean Mann, "Determining U.S. Commitments in Afghanistan," *Washington Quarterly*, Vol. 38, No. 1, 2015.

[39] Eggers, 2015.

APPENDIX

Details of the Expert Elicitation

The Delphi Method

The Delphi method was developed at RAND in the 1960s. While the technique has been refined over the years,[1] the fundamental premise remains the same. Experts individually make assessments or provide input and then offer written justification for those assessments. These experts are then given the opportunity to privately review the justifications offered by other participants and revise their assessments based on lines of reasoning that they had failed to include in their own initial calculations. The result is a consensual set of expert assessments based on more information than any single expert initially considered. Because participants work in private and remain anonymous to each other, final evaluations are reached without any of the psychological pitfalls of committee work, such as "specious persuasion, the unwillingness to abandon publicly expressed opinions, and the bandwagon effect of majority opinion."[2]

A simplified example of a Delphi exercise demonstrates this logic. Imagine that a Delphi exercise is convened to assure victory in a carnival game: The investigators wish to know how many peanuts are contained in a large glass pig. A panel of experts is assembled, including (among others) a physicist, a mathematician, a material scientist, a statistician, and an expert in the history of mountebanks.[3] Each performs his or her calculations and generates an estimate of the peanut content of the pig. Then, each is asked to justify his or her response, explaining the calculations involved. One participant might begin with the formula for the volume of an ellipsoid and then assume a volume for peanuts and proceed. Another might begin with the volume of an ellipsoid and then add a clever correction factor for the additional volume represented by the pig's feet and head. Yet another might simply use the volume of a sphere but add an innovative adjustment for the stochastic nature of the space between peanuts as they do or do not nest well with each other. The expert on mountebanks may not be able to articulate his or her volume calculation well at all but may make two critical observations about the kinds of tricks that carnival operators are likely to pull to make such estimation difficult—say, inconsistent thickness of the glass of the pig or the use of peanuts of different sizes. As the

[1] See, for example, Wong, 2003.

[2] Brown, 1968, p. 2.

[3] Commenting on the difficulty of staying abreast of developments in Afghanistan and amassing genuine expertise, a reviewer suggested that a more accurate analogy for the panel assembled for the carnival game would be "a kid who has played the game several times, a father who has watched the game being played many times and perhaps played himself as well, a graduate student who has studied carnival design, and the carnival worker operating the neighboring stand who thinks he understands how to beat the game based on conversations with the carnival worker actually running the game." We do not disagree.

experts review the justifications and calculations made by the others, they may recognize factors that they failed to include in their own calculations or come to understand that they have over- or underestimated some critical quantity. The revised estimates are likely to be based on more complex calculations, be better calculations, and be closer to each other than were the initial individual expert estimates.

The RAND Afghanistan Delphi Exercise

The RAND Afghanistan Delphi exercise was an iterative Delphi exercise based on the classic model. It was completed via iterative email exchange between March 11 and April 20, 2015. This section details the process used.

By definition, an expert elicitation is only as good as the experts elicited. An initial list of candidate participants was generated in consultation with senior RAND managers and based on participants in the previous exercises, documented in *Counterinsurgency Scorecard: Afghanistan in Early 2011 Relative to the Insurgencies of the Past 30 Years* and *Counterinsurgency Scorecard: Afghanistan in Early 2013 Relative to Insurgencies Since World War II*.[4] This list was expanded in consultation with the project sponsor and based on interactions in early briefings of draft results from *Paths to Victory: Lessons from Modern Insurgencies*.[5] An initial list of 25 candidates emerged from this process. Of this group, 14 initially agreed to participate. Of the 14 initial participants, all 14 completed the entire exercise. (In previous exercises, one or more of the initial participants failed to complete the entire sequence.) Participants included RAND staff with expertise on Afghanistan or who had deployed to Afghanistan; serving field-grade U.S. military officers with multiple (and recent) deployments to Afghanistan; military veterans who were current on COIN research; defense civilian representatives from the Office of the Secretary of Defense, the Joint Staff, and U.S. Central Command; government civilians with experience in Afghanistan; staff from the U.S. Agency for International Development; university faculty members; journalists; and experts from other prominent think tanks.

A summary set of instructions defined each factor more extensively and gave general guidance regarding the exercise. This guidance included the following text:

> Please evaluate each factor as present (1) or absent (0) in contemporary Afghanistan. All questions pertain to the area of conflict unless otherwise specified. If the answer is "well, it depends on where specifically you consider—that factor is present in part of the country but absent in another part," *please make "worst-case" assessments*. That is, if something is going well in RC [Regional Command] South but poorly in RC East, make your scoring based on RC East.
>
> A caveat: While we want "worst-case" assessments, we do not want assessments driven by isolated events. If a positive factor is present over the vast majority of the area of conflict, score it as present even if there are one or two isolated incidents where the factor was not present.

---

4 Paul, 2011; Paul, Clarke, Grill, and Dunigan, 2013a.

5 Paul, Clarke, Grill, and Dunigan, 2013b.

The iterative Delphi exercise included four scoring rounds with two phases in each round. Because it was clear that positions had become entrenched, and out of respect for participants' time, the second (discussion-only) phase of the third round was canceled by the moderator, progressing the exercise directly to the final phase. This resulted in a total of six phases. In the first phase of each round, participants provided scores for each factor, indicating whether they believed it was present ("1") or absent ("0") in Afghanistan. In the second phase of each round, participants were shown their scores relative to the mean scores of all participants. In all phases save the very first and the very last, participants were asked to justify their minority positions (factors on which they deviated from the group mean by 0.4 or more) and contribute to the ongoing discussion about the presence or absence of the factors.

In each phase save the first (nothing to discuss yet) and the last (discussion concluded), participants were asked to contribute to the discussion. In a traditional Delphi exercise, scorers are asked to justify all their ratings or calculations in the first round. However, because this exercise included 42 individual factors and because all participants volunteered their time, participants were asked to provide justification only for minority positions, lest a great quantity of volunteered time be consumed generating justifications for positions about which the entire panel was in complete agreement. In the second phase of each round, participants whose score on a factor differed from the group mean by 0.4 or more were informed that theirs was a minority position and asked to justify it. In this way, the discussion remained focused on factors that were actually contentious, rather than being diluted with justifications of factors about which there was already significant concordance. Scores that became minority positions in subsequent rounds (due to either changed scores or movement of the mean) were flagged as newly minority positions, indicating that a new justification was required from the participant.

After responding to requests for justification of their minority positions, participants were asked to weigh in on ongoing discussions of any of the factors. Space was made available for written rebuttals, counterarguments, endorsements, and so on, aimed at initial minority defenses or at discussion ensuing from them. No limit was placed on the volume or character of the discussion, though participants were encouraged to be concise. Instructions invited participants to refer to studies, data sets, personal experiences, or other evidence that they felt supported their positions or otherwise accounted for their reasoning or logic. The discussion was considerable, spanning more than 60 single-spaced pages in aggregate.

Raw Delphi Scores

Table A.1 presents the average score for each factor across the 14 participants who ultimately completed the exercise. Because the scoring was binary (0 or 1), the raw average can be accurately interpreted as the proportion of participants who indicated that a factor was present in their final scoring. For example, the first factor is "COIN force and government actions consistent with messages (delivering on promises)." The raw average is 0.43, which indicates that 43 percent of participants indicated that the factor was present (scored it as "1") in their final scoring.

For reference, Table A.1 also presents the consensus results (in the "Rounded Result" column), which were described in the main body of this report. Recall that 70-percent agreement was the threshold specified for consensus. This means that raw average scores of 0.7 or higher were considered to indicate consensus on presence and were rounded to "1" accordingly. Inversely, scores of 0.3 or lower were rounded to consensus absence, or "0." Scores between

Table A.1
Raw Average Scores from the RAND Afghanistan Delphi Exercise and Rounded Consensus Results

Factor	Raw Average	Rounded Result
COIN force and government actions consistent with messages (delivering on promises)	0.43	0.5
Forces of order maintains credibility with populations in the area of conflict (includes expectation management)	0.21	0
Messages/themes coherent with overall COIN approach	0.64	0.5
Forces of order avoids creating unattainable expectations	0.29	0
Themes and messages coordinated for all involved government agencies	0.07	0
Flow of cross-border insurgent support across significantly decreased, remains dramatically reduced, or largely absent	0	0
Important external support to insurgents significantly reduced	0.07	0
Important internal support to insurgents significantly reduced	0.21	0
Insurgents' ability to replenish resources significantly diminished	0	0
Insurgents unable to maintain or grow force size	0.07	0
COIN force efforts resulting in increased costs for insurgent processes	0.62	0.5
COIN forces effectively disrupt insurgent recruiting	0	0
COIN forces effectively disrupt insurgent materiel acquisition	0	0
COIN forces effectively disrupt insurgent intelligence	0.14	0
COIN forces effectively disrupt insurgent financing	0	0
Government leaders selected in a manner considered just and fair by majority of population in the area of conflict	0.46	0.5
Majority of citizens in the area of conflict view government as legitimate	0.43	0.5
Government corruption reduced/good governance increased since onset of conflict	0.29	0
Intelligence adequate to support kill/capture or engagements on COIN force's terms	0.86	1
Intelligence adequate to allow COIN force to disrupt insurgent processes or operations	0.93	1
COIN force of sufficient strength to force insurgents to fight as guerrillas	1	1
Unity of effort/unity of command maintained	0.64	0.5
COIN force avoids excessive collateral damage, disproportionate use of force, or other illegitimate applications of force	0.31	0.5
COIN force seeks to engage and establish positive relations with population in area of conflict	0.93	1
Short-term investments, improvements in infrastructure/development, or property reform in area of conflict controlled or claimed by COIN force	0.71	1
Majority of population in area of conflict supports/favors COIN forces	0.67	0.5
COIN force establishes and then expands secure areas	0.07	0
Government/COIN reconstruction/development sought/achieved improvements substantially above historical baseline	0.71	1
COIN force provides or ensures provision of basic services in areas it controls or claims to control	0.21	0
Perception of security created or maintained among population in areas COIN force claims to control	0.21	0
COIN force employs escalating repression	0	0
COIN force employs collective punishment	0	0
Government involves corrupt and arbitrary personalistic rule	0.86	1

Table A.1—Continued

Factor	Raw Average	Rounded Result
Country elites have perverse incentives to continue conflict	0.86	1
External professional military forces are engaged in fighting on behalf of the insurgents	0	0
Host nation is economically dependent on external supporters	1	1
Fighting is primarily initiated by the insurgents	0.77	1
COIN force fails to adapt to changes in adversary strategy, operations, or tactics	0.14	0
COIN force engages in more coercion/intimidation than insurgents	0.07	0
Insurgent force individually superior to COIN force by being either more professional or better motivated	0.43	0.5
Forces of order or allies rely on looting for sustainment	0.15	0
COIN force and government have different goals/levels of commitment	0.93	1

0.3 and 0.7 were left as contentious scores and rounded to 0.5, indicating a lack of agreement and neither presence nor absence.

Tables A.2 and A.3 show how these raw results on the 42 individual items roll up to produce the actual scorecard scores. Items indicated with numbers are scorecard factors and are either directly answered items or are composed of subfactors (listed with letters).

Table A.2
Good Factors Present in Early 2015 Afghanistan (Total: 7.5 of 15)

Good Factors	Subfactor	Factor
1. COIN force realizes at least two strategic communication factors (Score 1 if sum of a through e is at least 2)		0
a. COIN force and government actions consistent with messages (delivering on promises) (Score 1 if YES)	0.5	
b. COIN force maintains credibility with populations in the area of conflict (includes expectation management) (Score 1 if YES)	0	
c. Messages/themes coherent with overall COIN approach (Score 1 if YES)	0.5	
d. COIN force avoids creating unattainable expectations (Score 1 if YES)	0	
e. Themes and messages coordinated for all involved government agencies (Score 1 if YES)	0	
2. COIN force reduces at least three tangible support factors (Score 1 if sum of a through j is at least 3)		0
a. Flow of cross-border insurgent support significantly decreased, remains dramatically reduced, or largely absent (Score 1 if YES)	0	
b. Important external support to insurgents significantly reduced (Score 1 if YES)	0	
c. Important internal support to insurgents significantly reduced (Score 1 if YES)	0	
d. Insurgents' ability to replenish resources significantly diminished (Score 1 if YES)	0	

Table A.2—Continued

Good Factors	Subfactor	Factor
e. Insurgents unable to maintain or grow force size (Score 1 if YES)	0	
f. COIN force efforts resulting in increased costs for insurgent processes (Score 1 if YES)	0.5	
g. COIN forces effectively disrupt insurgent recruiting (Score 1 if YES)	0	
h. COIN forces effectively disrupt insurgent materiel acquisition (Score 1 if YES)	0	
i. COIN forces effectively disrupt insurgent intelligence (Score 1 if YES)	0	
j. COIN forces effectively disrupt insurgent financing (Score 1 if YES)	0	
3. Government realizes at least one government legitimacy factor (Score 1 if sum of a and b is at least 1)		1
a. Government leaders selected in a manner considered just and fair by majority of population in the area of conflict (Score 1 if YES)	0.5	
b. Majority of citizens in the area of conflict view government as legitimate (Score 1 if YES)	0.5	
4. Government corruption reduced/good governance increased since onset of conflict (Score 1 if YES)		0
5. COIN force realizes at least one intelligence factor (Score 1 if sum of a and b is at least 1)		1
a. Intelligence adequate to support kill/capture or engagements on COIN force's terms (Score 1 if YES)	1	
b. Intelligence adequate to allow COIN force to disrupt insurgent processes or operations (Score 1 if YES)	1	
6. COIN force of sufficient strength to force insurgents to fight as guerrillas (Score 1 if YES)		1
7. Unity of effort/unity of command maintained (Score 1 if YES)		0.5
8. COIN force avoids excessive collateral damage, disproportionate use of force, or other illegitimate applications of force (Score 1 if YES)		0.5
9. COIN force seeks to engage and establish positive relations with population in area of conflict (Score 1 if YES)		1
10. Short-term investments, improvements in infrastructure/development, or property reform in area of conflict controlled or claimed by COIN force (Score 1 if YES)		1
11. Majority of population in area of conflict supports/favors COIN forces (Score 1 if YES)		0.5
12. COIN force establishes and then expands secure areas (Score 1 if YES)		0
13. Government/COIN reconstruction/development sought/achieved improvements substantially above historical baseline (Score 1 if YES)		1
14. COIN force provides or ensures provision of basic services in areas it controls or claims to control (Score 1 if YES)		0
15. Perception of security created or maintained among population in areas COIN force claims to control (Score 1 if YES)		0

Table A.3
Bad Factors Present in Early 2015 Afghanistan (Total: 5.5 of 11)

Bad Factors	Subfactor	Factor
1. COIN force uses both collective punishment and escalating repression (Score 1 if sum of a and b is at least 1)		0
a. COIN force employs escalating repression (Score 1 if YES)	0	
b. COIN force employs collective punishment (Score 1 if YES)	0	
2. Government involves corrupt and arbitrary personalistic rule (Score 1 if YES)		1
3. Country elites have perverse incentives to continue conflict (Score 1 if YES)		1
4. External professional military engaged in fighting on behalf of insurgents (Score 1 if YES)		0
5. Host nation is economically dependent on external supporters (Score 1 if YES)		1
6. Fighting primarily initiated by the insurgents (Score 1 if YES)		1
7. COIN force fails to adapt to changes in adversary strategy, operations, or tactics (Score 1 if YES)		0
8. COIN force engages in more coercion/intimidation than insurgents (Score 1 if YES)		0
9. Insurgent force individually superior to COIN force by being either more professional or better motivated (Score 1 if YES)		0.5
10. COIN force or allies rely on looting for sustainment (Score 1 if YES)		0
11. COIN force and government have different goals/levels of commitment (Score 1 if YES)		1

Notes on Factors for Which Consensus Was Lacking

The expert panelists reached consensus on 33 of 42 individual scorecard items, leaving nine items on which they could not agree. Some of these factors on which consensus was lacking are particularly interesting.

One contentious factor was "COIN forces and government actions consistent with messages (delivering on promises)." According to our scoring, this factor actually improved from absent to contentious between 2013 and 2015. Since the Delphi exercises had some common members but were on the whole different panels of experts, there are likely to be slightly different views of the same factors. Still, there remains uncertainty about whether the change was due to the different composition of the panel or changing conditions in Afghanistan that led some experts to reach somewhat more optimistic views in 2015 than in previous years. The main area of disagreement in the discussion stemmed from the extent to which messages from the COIN forces and government actions were contradicted by corruption, contract extortion, and manipulation and the differences in capabilities between the national and local levels. While most panelists agreed that corruption was a major issue in delivering on promises, the majority believed that coherent and consistent messaging was still a work in progress under the Ghani administration but rhetoric and actions had been consistent with regard to the ongoing insurgency, reconciliation efforts with the Taliban, and Afghanistan's relationship with the broader international community. More recently, concerns over the spread of Islamic State fighters into Afghanistan have prompted Ghani to deliver a clear and consistent message that the group will be countered wherever its presence is detected.[6] Still, defeating the Islamic State in South Asia will be difficult, as the group (which calls its South Asia franchise the Islamic State of Khorasan) attempts to co-opt disaffected local militants and challenge both the Taliban and al Qaeda throughout the region.[7]

"Messages/themes coherent with overall security approach" was another disputed factor, as it was in 2013. Participants who saw this factor as absent offered a range of critiques. First, some remarked that operations conducted by the security forces were politicized, undertaken only to benefit the image of the government and not to provide meaningful security for the Afghan people. Second, despite being lauded as the first line of defense, the Afghan National Police remained largely incompetent, poorly motivated, corrupt, and unable to protect the population. Finally, some noted that the transition from Karzai to Ghani made articulating a consistent message problematic, though others found this to be less of an issue and something that would likely resolve itself over time. The recent mobilization of militias and warlords in northern Afghanistan to fight the Taliban has also been perceived as directly contradicting messages put forth by the Afghan government that the security forces are "holding their own" in the fight against the insurgents.[8] Delphi participants who found this factor to be present argued that Afghan government and security force communications corresponded to the overall strategy and that these entities had done a praiseworthy job in minimizing the "say-do" gap. Overall, there is a fair amount of consistency between how the government describes the campaigns it executes and how they are executed in practice. While the objectives might be slightly overreached, or the scope or impact somewhat exaggerated, the messages clearly meet the threshold of "coherent."

Another contentious factor was "COIN force efforts resulting in increased costs for insurgent processes," which went from agreed upon as present in 2013 to disputed in 2015. Most of the debate revolved around whether or not the ANSF has shifted to a more defensive posture over the past year, which some argued was the result of changed authorities, fewer resources, and a generally slower overall operational tempo. With less pressure, the cost to the insurgents for maintaining various processes has decreased, as the Taliban now have greater freedom of movement and more access to illicit revenue streams. Another panelist replied that while the Afghan security forces grow more competent each year, on balance, for the past two years they have been largely static and reactive and rarely take the fight to the enemy. On the other hand, participants who found this factor to be present argued passionately that while the conventional corps of the ANSF might be reactive or defensive in some areas, the security forces also undertook several major operations in early 2015 involving conventional, special, and Ministry of Interior forces that successfully curtailed insurgent supply networks and facilitation routes—actions that almost certainly hindered the insurgents during the 2015 fighting season.

---

[6] For more, see Borhan Osman, "The Shadows of 'Islamic State' in Afghanistan: What Threat Does It Hold?" Afghanistan Analysts Network, February 12, 2015; see also, Rebecca Zimmerman, "Has Islamic State Entered Afghanistan?" *World Affairs*, April 29, 2015.

[7] Seth G. Jones, "Expanding the Caliphate," *Foreign Affairs*, June 11, 2015.

[8] Mujib Mashal, Joseph Goldstein, and Jawad Sukhanyar, "Afghans Form Militias and Call on Warlords to Battle Taliban," *New York Times*, May 24, 2015.

"Government leaders selected in a manner considered just and fair by majority of population in the area of conflict" was another contentious factor, but one that moved in a positive direction. This move from absent to no consensus ("0" to "0.5") was largely due to the election of a new president, Ashraf Ghani. According to several expert panelists, 2014's presidential and legislative elections were considered fair, while the rejection of cabinet nominees by the parliament demonstrated that it was not a "rubber stamp" for the executive. Another expert pointed to the fact that local residents in contested areas pressured their lower-level Islamic Emirate of Afghanistan leadership to permit them to vote or the leadership actually encouraged constituents to vote as evidence that the overall selection process was viewed as fair. Experts who found this factor absent were skeptical that the majority of the population in the area of conflict—especially the predominantly Pashtun areas in the south and east—felt that Ghani was elected in a just and fair manner. Some of this probably has to do with overall cynicism about democracy in the first place, or with Pashtun claims that Ghani was handpicked by the West to succeed Karzai because Ghani is perceived as more pliable. Although Ghani is Pashtun, he spent a considerable amount of time living in the United States and thus is likely viewed with distrust by many Afghans, according to these views.

Related was disagreement about whether the majority of citizens view the Afghan government as legitimate in the area of conflict. The new national unity government has brought a renewed feeling of confidence in the potential for better governance. Although the election and subsequent stalemate became contentious, Ghani and Abdullah are generally viewed as legitimate. However, the strength of this claim diminishes when one considers areas of conflict where, clearly, some Afghans support the insurgency in opposition to the government. Moreover, 2014 saw demonstrable action by Taliban shadow governments in these areas, reflecting concentrations of popular resistance activity and growing support for the insurgency.

The factor "Unity of effort/unity of command [of COIN forces] maintained" was contentious but also improved since 2013, when it was scored as absent. Those who found this factor problematic in 2015 cited such issues as the apparent divides among the Afghan National Army, the Afghan National Police, the Afghan Local Police, and some of the more ad hoc community/village militias (with the militias considered the most problematic) that are relied on to provide local security. Participants noted that coordination within and between these separate entities remained a challenge, and the insurgents were able to regularly exploit these seams. Another participant was more optimistic, admitting that while there were some exploitable seams between Afghanistan's Ministry of Interior and Ministry of Defense in certain localities, on the whole, the ANSF is working together extremely well with law enforcement—from conducting joint operations to sharing intelligence. Although the National Directorate for Security was not mentioned specifically, the directorate does claim some jurisdiction over the Afghan Local Police as part of its domestic security responsibility.[9]

A contentious issue in and of itself, "COIN force avoids excessive collateral damage, disproportionate use of force, or other illegitimate applications of force" is an issue that has been frequently and passionately debated among numerous parties to the conflict, including ISAF forces, Afghan security forces, Afghan civilians, and the insurgents themselves (who often make false claims of COIN force collateral damage in an attempt to bolster their propaganda efforts). Moreover, media outlets and nongovernmental organizations have joined the debate,

---

[9] For more details, see International Crisis Group, *The Future of the Afghan Local Police*, Asia Report 268, Kabul and Brussels, June 4, 2015, p. 9.

some with less-than-reliable data and thinly veiled political motives. There are disagreements about the number of civilian casualties and also about which side (the COIN forces or the insurgents) is predominantly responsible for those deaths. One panelist referenced the United Nations Assistance Mission in Afghanistan civilian casualty report from February 2015, which indicated a significant increase in civilian casualties from ground engagement between anti-government forces and the ANSF throughout all regions of Afghanistan.[10]

One prominent rebuttal to this point was offered by another panelist, who said that measuring this factor was difficult for many reasons; specifically, there is often a significant difference between UNAMA civilian casualty numbers and the statistics compiled by ISAF, an organization with a fairly mature investigative, validation, and vetting process for such allegations. Yet, perhaps the most poignant remark insisted that, unlike providing basic services, this is a factor for which just trying does not count for very much. To be sure, there is a major difference between attempting to avoid collateral damage and actually avoiding said damage. Realistically, Afghans are likely upset with both the Taliban and the COIN forces. The Taliban are seen as brutal but capable of swift and fair justice. The COIN forces are seen as trying to provide security but often corrupt in how they do so.

In 2013, Delphi exercise participants scored "Majority of population in areas of conflict supports/favors COIN forces over the insurgents" as present, but this same factor was contested in 2015. What changed? While some of the difference can be accounted for by differences of opinion between different groups of Delphi experts, one panelist correlated a decrease in support for the COIN force with an increase in attacks by the insurgents that took the form of massed forces. In essence, if the majority of the population in areas of conflict supported or favored the COIN force over the insurgents, there would be far fewer attacks by the Taliban, especially these types of attacks, according to one subject-matter expert. The expert went on to note that where the insurgents were operating with such impunity that they could effectively mass large formations, the population would likely alert security officials if it truly supported the COIN effort, even at a risk of insurgent reprisal. One participant countered that this notification was, in fact, happening, with tip lines receiving an increasing number of calls and Afghan security officials otherwise receiving information from locals who want insurgents removed. The debate over this factor did highlight an important point, which is that insurgency supporters are probably less likely to talk to journalists, especially Western journalists. Moreover, although polling data (almost universally acknowledged as less than perfect) unequivocally suggests that support for the ANSF is lower than the national average in areas of conflict, it also remains well above the level of professed support for the insurgents in most (but not all) of these areas.[11]

"Insurgent force individually superior to COIN force by being either more professional or better motivated" was the only bad factor that was contentious, and it was also contentious on the 2013 scorecard. Despite years of training and resources provided by ISAF, there were still significant reports in early 2015 of low morale, ghost attendance, and insufficient equipment in the ANSF. Some expert panelists seemed to split the difference on this factor, noting that while the ANSF might be more professional, the insurgents were superior because of their motivation—driven by a combination of variables from religion to resistance in the face of a

---

[10] United Nations Mission in Afghanistan, *Protection of Civilians in Armed Conflict: Afghanistan Annual Report 2014*, February 2015.

[11] Zach Warren, *Afghanistan in 2014: A Survey of the Afghan People*, San Francisco, Calif.: Asia Foundation, 2014.

perceived occupation. One point that was raised several times was that the insurgents were far less well equipped than the ANSF and have been able to remain a viable fighting force despite the threat of overwhelming U.S. and coalition firepower, even if attempts to increase the cost of the insurgency have been somewhat successful.

References

Abrahms, Max, "'Mohammad Omar's Death Could Help the Afghan Peace Process—or Harm It," *Washington Post*, *Monkey Cage* blog, August 7, 2015. As of October 30, 2015: https://www.washingtonpost.com/blogs/monkey-cage/wp/2015/08/07/will-mullah-omars-death-benefit-or-harm-the-peace-process/

Ahmed, Azam, and Joseph Goldstein, "Taliban Gains Pull U.S. Units Back into Fight in Afghanistan," *New York Times*, April 29, 2015.

Blank, Jonah, "Give Ghani a Chance: Why This Time Is Different," *Foreign Affairs*, March 31, 2015.

Brown, Bernice B., *Delphi Process: A Methodology Used for the Elicitation of Opinions of Experts*, Santa Monica, Calif.: RAND Corporation, P-3925, 1968. As of October 30, 2015: http://www.rand.org/pubs/papers/P3925.html

Cordesman, Anthony H., *The Afghan Campaign and the Death of Mullah Omar*, Washington, D.C.: Center for Strategic and International Studies, August 2, 2015.

Craig, Tim, "Afghan Forces Straining to Keep the Expanding Taliban at Bay," *Washington Post*, May 16, 2015.

Clarke, Colin P., and Christopher Paul, *From Stalemate to Settlement: Lessons for Afghanistan from Historical Insurgencies That Have Been Resolved Through Negotiations*, Santa Monica, Calif.: RAND Corporation, RR-469-OSD, 2014. As of October 30, 2015: http://www.rand.org/pubs/research_reports/RR469.html

Dobbins, James, and Carter Malkasian, "Time to Negotiate in Afghanistan," *Foreign Affairs*, July–August 2015, pp. 62–63.

Eggers, Jeff, "Afghanistan, Choose Your Enemies Wisely," *Foreign Policy*, August 24, 2015.

Goldstein, Joseph, and Taimoor Shah, "Death of Mullah Omar Exposes Divisions Within Taliban," *New York Times*, July 30, 2015.

Houreld, Katharine, "Pakistan Military Says Its Spies Will Cooperate with Afghanistan," Reuters, May 19, 2015.

International Crisis Group, *The Future of the Afghan Local Police*, Asia Report 268, Kabul and Brussels, June 4, 2015.

Jones, Seth G., "Expanding the Caliphate," *Foreign Affairs*, June 11, 2015.

Malkasian, Carter, "Is Peace Possible in Afghanistan?" *Foreign Affairs*, August 18, 2015.

Mashal, Mujib, Joseph Goldstein, and Jawad Sukhanyar, "Afghans Form Militias and Call on Warlords to Battle Taliban," *New York Times*, May 24, 2015.

Michaels, Jim, "Decade of War, Billions in U.S. Aid Fail to Defeat Taliban," *USA Today*, May 19, 2015.

Osman, Borham, "The Shadows of 'Islamic State' in Afghanistan: What Threat Does It Hold?" Afghanistan Analysts Network, February 12, 2015. As of October 30, 2015: https://www.afghanistan-analysts.org/the-shadows-of-islamic-state-in-afghanistan-what-threat-does-it-hold

Paul, Christopher, *Counterinsurgency Scorecard: Afghanistan in Early 2011 Relative to the Insurgencies of the Past 30 Years*, Santa Monica, Calif.: RAND Corporation, OP-337-OSD, 2011. As of October 30, 2015: http://www.rand.org/pubs/occasional_papers/OP337.html

Paul, Christopher, Colin P. Clarke, and Beth Grill, *Victory Has a Thousand Fathers: Sources of Success in Counterinsurgency*, Santa Monica, Calif.: RAND Corporation, MG-964-OSD, 2010. As of October 30, 2015: http://www.rand.org/pubs/monographs/MG964.html

Paul, Christopher, Colin P. Clarke, Beth Grill, and Molly Dunigan, *Counterinsurgency Scorecard: Afghanistan in Early 2013 Relative to Insurgencies Since World War II*, Santa Monica, Calif.: RAND Corporation, RR-396-OSD, 2013a. As of October 30, 2015: http://www.rand.org/pubs/research_reports/RR396.html

———, *Paths to Victory: Lessons from Modern Insurgencies*, Santa Monica, Calif.: RAND Corporation, RR-291/1-OSD, 2013b. As of October 30, 2015: http://www.rand.org/pubs/research_reports/RR291z1.html

Paul, Christopher, Jennifer D. P. Moroney, Beth Grill, Colin P. Clarke, Lisa Saum-Manning, Heather Peterson, and Brian Gordon, *What Works Best When Building Partner Capacity in Challenging Contexts?* RR-397-OSD, Santa Monica, Calif.: RAND Corporation, 2015. As of October 30, 2015: http://www.rand.org/pubs/research_reports/RR937.html

Raghavan, Sudarasan, "As the U.S. Mission Winds Down, Afghan Insurgency Grows More Complex," *Washington Post*, February 13, 2015.

———, "Foreign Fighters Loyal to ISIS are Now Flowing into Afghanistan," *Washington Post*, April 14, 2015.

Shinn, James, and James Dobbins, *Afghan Peace Talks: A Primer*, Santa Monica, Calif.: RAND Corporation, MG-1131-RC, 2011. As of October 30, 2015: http://www.rand.org/pubs/monographs/MG1131.html

United Nations Mission in Afghanistan, *Protection of Civilians in Armed Conflict: Afghanistan Annual Report 2014*, February 2015. As of October 30, 2015: http://unama.unmissions.org/Portals/UNAMA/human%20rights/2015/2014-Annual-Report-on-Protection-of-Civilians-Final.pdf

U.S. Department of Defense, *Progress Toward Security and Stability in Afghanistan*, Washington, D.C., October 2014. As of October 30, 2015: http://www.defense.gov/Portals/1/Documents/pubs/Oct2014_Report_Final.pdf

Warren, Zach, *Afghanistan in 2014: A Survey of the Afghan People*, San Francisco, Calif.: Asia Foundation, 2014. As of October 30, 2015: https://asiafoundation.org/resources/pdfs/Afghanistanin2014final.pdf

Watts, Stephen, and Sean Mann, "Determining U.S. Commitments in Afghanistan," *Washington Quarterly*, Vol. 38, No. 1, Spring 2015, pp. 107–124.

Wong, Carolyn, *How Will the e-Explosion Affect How We Do Research? Phase I: The E-DEL+I Proof-of-Concept Exercise*, Santa Monica, Calif.: RAND Corporation, DB-399-RC, 2003. As of October 30, 2015: http://www.rand.org/pubs/documented_briefings/DB399.html

Wong, Edward, and Mujib Mashal, "Taliban and Afghan Peace Officials Have Secret Talks in China," *New York Times*, May 25, 2015.

Zimmerman, Rebecca, "Has Islamic State Entered Afghanistan?" *World Affairs*, April 29, 2015.